Releasing the Commons

This book moves beyond seeing the commons in the past tense, an entity passed over from the public into the private, to reimagine the commons as a process, a contest of force, a reconstitution, and a site of convening practices. It highlights new spaces of gathering opening up, such as the digital commons, and new practices of being in common, such as community economies and solidarity networks. The commons is seen as a contested domain of the collective and as a changing way of being in common, with the balance poised in the tensile play between political economy and social innovation. The book focuses on the possibility of recovering a future in which more can be held by the many, focusing on three concepts: nation and nature as a commons, publics and rights, and bodies, concerning the management of lives and livelihoods. Across these three passage points, the book finds evidence of a commons under attack but also defended in fragile though promising ways.

With contributions from leading scholars, this thought-provoking book will be of great interest to students and scholars in geography, environmental studies, politics, anthropology, and cultural studies.

Ash Amin is Professor of Geography in the Department of Geography at the University of Cambridge, UK.

Philip Howell is Senior Lecturer in the Department of Geography at the University of Cambridge, UK.

Routledge Studies in Human Geography

This series provides a forum for innovative, vibrant, and critical debate within Human Geography. Titles will reflect the wealth of research which is taking place in this diverse and ever-expanding field. Contributions will be drawn from the main sub-disciplines and from innovative areas of work which have no particular sub-disciplinary allegiances.

For a complete list of titles in this series, please visit www.routledge.com/series/ SE0514.

Releasing the Commons
Rethinking the futures of the commons

Edited by
Ash Amin and
Philip Howell

LONDON AND NEW YORK

First published 2016
by Routledge
2 Park Square, Milton Park, Abingdon, Oxon OX14 4RN

and by Routledge
711 Third Avenue, New York, NY 10017

First issued in paperback 2018

Routledge is an imprint of the Taylor & Francis Group, an informa business

British Library Cataloguing-in-Publication Data
A catalogue record for this book is available from the British Library

Library of Congress Cataloging-in-Publication Data
Names: Amin, Ash, editor. | Howell, Philip, 1965– editor.
Title: Releasing the commons : rethinking the futures of the commons /
edited by Ash Amin and Philip Howell.
Description: Abingdon, Oxon ; New York, NY : Routledge, 2016. | Series:
Routledge studies in human geography
Identifiers: LCCN 2015047064| ISBN 9781138942349 (hardback) |
ISBN 9781315673172 (e-book)
Subjects: LCSH: Common good–International cooperation. | Commons–
Political aspects. | Social policy. | Cultural geography. | Political sociology. |
Political geography.
Classification: LCC JC330.15 .R46 2016 | DDC 333.2–dc23
LC record available at http://lccn.loc.gov/2015047064

ISBN 13: 978-1-138-54648-6 (pbk)
ISBN 13: 978-1-138-94234-9 (hbk)

Typeset in Times New Roman
by Wearset Ltd, Boldon, Tyne and Wear

Contents

vi *Contents*

Figures

Tables

Contributors

Ash Amin is Professor of Geography in the Department of Geography at the University of Cambridge, UK.

Nicholas Blomley is Professor of Geography at Simon Fraser University, Canada. He has a long-standing interest in the geographic dimensions of law, with a particular focus on issues regarding property in land.

Jenny Cameron is an Associate Professor in the Geography and Environmental Studies Discipline at the University of Newcastle, Australia. In 2015, she was a Visiting Professor at the Chinese University of Hong Kong. Her research focuses on forms of economic subjectivity and collective action associated with diverse and community economies.

Renu Desai is the Coordinator of the Centre for Urban Equity, CEPT University, Ahmedabad, India. Her research examines urban informality and urban transformation in Indian cities, with a focus on questions of equitable development and urban citizenship. She is co-editor of *Urbanizing Citizenship: Contested Spaces in Indian Cities* (Sage, 2012).

Maria Fannin is Senior Lecturer in Human Geography at the University of Bristol, UK. Her research interests include the social and economic implications of bioscience research, and specifically the collection and exchange of human biological materials. Her work focuses on new conceptualisations of labour and exchange in the global (bio)economy, and on feminist geographical approaches to a 'bodily commons' in a post-genomic age.

Natalie Fenton is a Professor in Media and Communications at Goldsmiths, University of London, UK, where she is Co-Director of the Goldsmiths Leverhulme Media Research Centre and Co-Director of Goldsmiths Centre for the Study of Global Media and Democracy. Her latest book (Polity, 2016) will be *Left Out?: New Media and Radical Politics*.

J.K. Gibson-Graham is the pen-name shared by feminist economic geographers Katherine Gibson and the late Julie Graham. Katherine Gibson is a Professorial Research Fellow at the Institute for Culture and Society, Western Sydney University, Australia. J.K. Gibson-Graham founded the Community Economies

Collective (see www.communityeconomies.org) and the international Community Economies Research Network whose members share an interest in theorising, representing, and ultimately enacting new visions of economy.

Stephen Healy has a PhD in Geography from the University of Massachusetts and is a Research Fellow at the Institute for Culture and Society, Western Sydney University, Australia. He is a series editor for the Diverse Economies and Liveable Worlds book series with University of Minnesota Press and an Associate Editor for the journal *Rethinking Marxism*. His research uses insights from psychoanalytic and Marxian theory to trace connections between subjectivity and economy.

Philip Howell is a Senior Lecturer in the Department of Geography, University of Cambridge, UK, working principally on the historical geography of social regulation in nineteenth-century Britain. He has published monographs on the regulation of prostitution in Victorian Britain and the British Empire, and on the dog's place in the cultural landscape of Victorian Britain.

Alex Jeffrey is a Reader in Human Geography at the Department of Geography in the University of Cambridge, UK. His work explores the fraught processes through which coherent ideas of the state are communicated across space in the wake of violent conflict. In recent work he has examined the spatial implications of war crimes trials, looking in particular at the material and embodied processes through which the legitimacy of law is communicated, resisted, and reworked.

Bruce Lankford is Professor of Water and Irrigation Policy in the School of International Development at the University of East Anglia, UK. He has worked for more than 30 years in the fields of irrigation and water resources management, starting in Swaziland in 1983. His main research and advisory work covers water management in sub-Saharan Africa. He is a co-founder of the UEA Water Security Research Centre and a Fellow of the Institution of Civil Engineers.

Colin McFarlane is Reader in Urban Geography at Durham University, UK. His work examines the experience and politics of informal settlements, and has included work in Mumbai, Kampala, and Cape Town. He is the author of *Learning the City: Knowledge and Translocal Assemblage* (Blackwell, 2011) and co-editor of several books, including *Smart Urbanism: Utopia Vision or False Dawn?* (with Simon Marvin and Andres Luque, Routledge, 2016) and *Infrastructural Lives* (with Steve Graham, Routledge, 2015).

Sarah A. Radcliffe works on development, social inequality, and citizen-state relations in Andean Latin America, from decolonial and feminist perspectives. Her recent work has focused on indigenous women's critical knowledge production about development interventions. Her latest book is *Dilemmas of Difference: Indigenous Women and the Limits of Postcolonial Development Policy* (Duke University Press, 2015).

Adam Reed is a Senior Lecturer in Social Anthropology at the University of St. Andrews, UK. His research includes fieldwork conducted in Papua New Guinea and the United Kingdom, and ranges between legal anthropology, anthropology of ethics, anthropology of the city, and anthropology and literature.

Marilyn Strathern, former William Wyse Professor of Social Anthropology at Cambridge University, is (honorary) life president of the Association of Social Anthropologists (UK and Commonwealth). She recently chaired a working party on organ donation for the Nuffield Council on Bioethics.

John Urry was Distinguished Professor of Sociology, and Co-Director of the Institute for Social Futures, Lancaster University, UK. John sadly passed away as this book was going to press; his final book, What is the Future? will be published by Polity in 2016.

Acknowledgements

This book is the outcome, appropriately enough, of a collective initiative. This was begun by Ash Amin and Kendra Strauss in the Department of Geography, University of Cambridge, and resulted, with Ash Amin and Philip Howell taking the lead, in an international, interdisciplinary symposium taking the title 'The Shrinking Commons', which was held in Cambridge on 8 and 9 September 2014. We are immensely grateful to everyone who participated in this symposium, not only those represented directly in this volume but also the speakers who also made this such a special event, namely: Alun Anderson, Filip de Boeck, Eduardo Mendieta, Denise Morado, Henrietta Palmer, Edgar Pieterse, Keith Richards, Sarah Whatmore, and Kathryn Yusoff. The demands of publication and on individual participants mean that we are not able to include contributions from everyone who was involved, but this meeting would not have succeeded as it did, or developed as it has, were it not for the collective interest and enthusiasm of a much larger community. We would like to take this opportunity, formally but no less warmly, to thank all the participants in this symposium for their generosity and collegiality.

We are very grateful to the Department of Geography for providing financial support for the symposium, support that was supplemented by a contribution from Emmanuel College, Cambridge. This event simply would not have been possible at all were it not for the efforts of Bill Adams, Danielle Feger, and Urša Mali, and we would also like to thank Helen Pallett and Dave McLaughlin for their help.

Shortly before publication we heard the sad news of John Urry's untimely death. In addition to a distinguished academic career, John was a tireless campaigner for justice, and a true friend and colleague. We are very sorry that he will not see this work published, and would like to dedicate this book to him.

1 Thinking the commons

Ash Amin and Philip Howell

Introduction

The commons have been described as a drama, even – most famously – a 'tragedy'. Their fate, their future, has never seemed more parlous, with climate change, population growth, and competition for scarce resources seemingly threatening our greatest common property, the planet itself. Enclosure – once seen as the very end of the commons – is touted by some as the only practicable way to protect precious environments subject to the existential threat of encroachment. At the same time, undergirded by a general anxiety that the natural, social, and political commons are at risk from capitalism's bloating expansion, from hyper-consumption, and from corporatist politics, critics of enclosure grasp at a new narrative, proposing a new global commons as the only solution to our pressing global problems. Far from moribund, then, and affronting attempts to reduce their purchase to the mere status of 'public goods', the commons remain central to the material struggles and imaginaries of collective well-being, now and into the near future. The commons is dead: long live the commons!

Yet how should we understand these contemporary conjugations of the commons, if by this term we understand a process, a contest of force, a reconstitution, a site of convening practices? Our justification for this volume is that there is a pressing need to reconceive the 'commons', to think of the commons as more than a tug of war between use value and exchange value, between common use and commodification, between communities and corporations, in which the odds are always and everywhere stacked against the continuing existence and vitality of the commons. Without ignoring the facts of the systematic encroachment on life, resources, and spaces once held in common, at the same time we envisage the opening up of new spaces of cooperation and collective action, such as the digital commons, new practices simply of 'being in common', community economies and solidarity networks. We see the contemporary commons as both being lost in old shapes and recovered in new forms, as, in brief, a contested and dynamic domain of collective existence, with the balance delicately poised between the rapacious demands of political economy and the promise of social innovation.

This play of forces, at a moment of collective crisis, is one focus of this book, which remains open to diverse meanings of the commons, ranging from rules of collective return to new practices of association, precisely in order to nurture and expand the politics of possibility in our seemingly hyper-privatised age. The contributions to this collection remain soberly aware of the problems posed to public and collective ownership, but at the same time potentially adjustable through new practices of shoring and stewardship. They explore the shifts that are inherent in any 'being in common', seeing old institutions such as law and citizenship compromised as sites of universal inclusion, but noting also the rise of new spaces supporting both secessionist and collectivist ambitions, as well as micro-worlds of social being born out of the ruins of neoliberal abandonment. They find, notably, a vibrant counter-culture of non-utilitarian living and associating, surviving and even flourishing amid the detailed disciplinary calculations of market society, and often couched in the language of protecting the shrinking commons.

To recover the commons, then, may be a matter of discursive framing as well as or as much as a political event. Thought of as a rule or possession, the future of the commons for sure seems bleak. Things once held free or at least public are becoming ever more securely privatised and commoditised, typically against any reasonable measure of the long-term general interest, for all that the states, corporations, and elites of global capitalist society protest otherwise. The world over, the natural world, the global reservoir of pooled resources, human labour in all its forms, mutually-beneficial commerce, law, public policy, culture, and knowledge, the body and the bodily realm, and even life itself, are up for grabs, violently opened up to neoliberalism's rigged markets and closed down as the spaces and means of collective stewardship. The language of government in the public interest has itself begun to sound anachronistic and implausible, however necessary its revival, as several chapters in this book argue. The commons no longer seems to be part of the common sense of political economy, reduced rather to a faint echo of the moral economy of the world we have lost. In contrast, however, if we think of the commons as a practice or process, the future looks less dismal, as is also increasingly recognised. Experiments the world over in land cultivation and conservation, micro-enterprise and labour, ethical trade and responsible consumption, community orientation and social networking, and participatory and democratic politics, provide ample evidence of the continuing social interest in collaboration, collective orientation, and future stewardship. Discrete and diverse as are these examples of the lively lived spaces of actually existing commons (Eizenberg 2012), they intimate the possibility of an expanded politics of the commons, one that can build on practice to lever a change in the rules and principles of possession.

Thus, to those who would still object that the commons is a lost cause, and that the use of any such language is a gesture of romantic fancy, we would argue that by thinking the commons anew, all the time recognising their inherent ambiguities and latent complexities, a persuasive politics for nature and a truly shared society can be mounted. This book is neither a commonist manifesto nor a call-to-arms, though we share the same concerns of those that already exist.[1] We aim, rather, at a collective, cooperative audit of transformations, driving forces, and commoning experiments, pressing towards outlining the new possibilities for the commons. Our contributors and collaborators share the animus of those who have sought to pit the promise of commoning against the forces of commodification, and of those who reject the blandishments of a global political and economic order that has enclosed the commons in the politics of nostalgia or fantasy. We are all keenly aware of the critical damage that privatisation and the careering of a rampant exchange value has done and is doing, to an increasingly vulnerable world. As far as the commodity form goes, a seemingly ever more entrenched neoliberalism has cast aside all that cannot be privately owned or tended as anachronistic and anomalous. And how characteristic is it that the coordinated attacks on the sustainability of even modest lives, widely understood to be the hallmarked product of corporatist capitalism, are met with ever more rounds of deregulation, privatisation, and dispossession.

There are grounds enough for anger and indignation, as shown by many of this book's contributors: the artful practices of 'offshoring' that threaten to undermine national responsibility (John Urry), the coalitions of law, language, government, and elitism making deep inroads into access to land and livelihoods (Nick Blomley, Alex Jeffrey, Colin McFarlane, Renu Desai, and Sarah Radcliffe), to autonomy and an effective democracy, and indeed in response to the comprehensive erosion of all things held in common (Natalie Fenton, J.K. Gibson-Graham, Jenny Cameron, and Stephen Healy). The mass production of misery and insecurity unleashed by these *expulsions*, as Saskia Sassen (2014) calls them – of nature from itself, land from future or common use, people from employment, housing and welfare, money from the productive economy, growth from its spatial and political communities, citizens from the polity, and states from their peoples – are rampant and unrelenting. They persist virtually unopposed in an age of political as well as economic austerity, justified by the apologists of neoliberalism as necessary steps towards a better future.

If there is for us a single icon of these *shrinking* commons, it is the darkening of the Arctic ice cap as anthropogenic global warming has led to the retreat of the summer sea ice – a process that nevertheless reveals new opportunities for global capitalism and its masters (Anderson 2009). The fate of the Arctic speaks of a future of physical, material spaces no longer safe anywhere in the world from the shadow of the commodity and corporate enclosure. And yet it remains true that in spite or because of these encroachments, the consciousness of a collective common life endures, of the planetary precarity that affects us all, and all forms of life. The commons thus spring back as an inevitable co-product of the forces that beset the world, their survival – in desire and in practice (Williams 2005) – an ineradicable condition of our necessarily collective existence. This is echoed in Michael Hardt and Antonio Negri's recognition of what they call 'the common' rather than the commons (2006: xv) in a world that no longer has an 'outside': 'In the era of globalization, issues of the maintenance, production, and distribution of the common in both these senses and in both ecological and socioeconomic frameworks become increasingly central' (Hardt and Negri 2009: viii). Others, who are rather less convinced by claims regarding the ontological necessity of 'the common', or the pre-eminence of biopolitical labour and the mass 'cognitivisation' of knowledge capitalism, still readily accept that 'new enclosures have demonstrated that not only has the common not vanished, but that new forms of social cooperation are constantly being produced' (Federici 2010: 284). Of course we should not overplay this hope-filled, affirmative stance, let alone fall into complacency at the ineradicability and impressive fecundity of 'the common', but if such commentators are right, it might be premature to write-off the commons, and their value and necessity. The question, then, becomes one of how the commons should be spoken of, and, just as importantly, how they may be made politically resonant. Is it the noun or the verb in the commons that provides the opportunity? Should a politics of the commons focus on things held in reserve or in public, on collaboration among humans and with nature, on the terms and terrains on which decisions about the collective future are made?

Things and practices

These options are not mutually exclusive, for the commons (to revisit its etymology) are notably *commodious*, in the sense of being both spacious and timely.[2] They offer diverse opportunities and inspirations for political action, and in this collection we open a window onto their nature and the possibilities they offer for our time and our world. The political urgency of these questions is clear from any audit of this kind. It is clear from the encroachments revealed by our own contributors that (at the very least) we are back to having to make a strong case for the common or collective in the form of public ownership, shared goods and services, social equity, universal rights, protected nature, and shared cities, however much we may need to explore the necessary nuances.[3] Readers will see that these encroachments are by no means unambiguous, involving as they do, unstable mixes of commodified and altruistic personhood, double-edged mobilisations of law, the hollowing out and restitution of nation and nature, forms of closure and openness in new communication spaces, and both elite capture and grass-root response. The spaces of encroachment are typically sites of struggle and imaginative commoning. There are signal examples in this book of new digital affordances, nature curated with care by communities, push-backs against privatisation and corporate encroachment, imaginative forms of urban improvisation, and social organisation based on reciprocity and empathy. These practices disturb the singularity of appropriation and expulsion, by making commoning a way of being in the world, as Rebecca Solnit (2001, 2006) would have it.

This is not to say that the battle between encroachment and commoning is a duel among equals. Far from it, and in fact there is no unambiguous sign yet of the contradictions of encroachment, such as the alienation of nature and society from itself, giving way to a new, collaborative order on the kind of scale that is clearly necessary. We have to be suspicious, as Gibson-Graham (1996) advises, of proclamations of the end of capitalism and the inevitable rise of the communal or common; waiting for the gravediggers of capitalism is always like waiting for Godot. If anything, the evidence from the stress fields across the world is not encouraging. There is no significant push back in the Americas, Africa, Europe, Asia, and the Middle East against the causes and beneficiaries of climate change, neoliberal inequality, corporatism, financial speculation, capital flight, and oligarchy and elitism. Resistance is quickly swept aside, made fragile, relegated to a parallel margin. Democracy itself, at least as practised in the neoliberal heartlands, has been made complicit with the ballot skewed by public disengagement, collusion between the media, corporations and the state, and powerful status quo narratives preying on fear and anxiety. So, for instance, migration and the question of the free movement of labour, not to mention the rights of refugees and exiles, have become, at various levels, a source of grievances over entitlement to shrinking common-pool resources. The politics of the commons, in this respect, may be mobilised in the service of neoliberalism and neoconservatism. This of course harks back to the narrative of the tragedy of the commons. But this complicity plays, furthermore, on the curiously shared etymology between commodity and commons.[4] The commodity has a notable 'double nature': the propriety and convenience of its commodious use value, but at the same time its fungibility, its exchange value (Linebaugh 2008: 63). The commodity, just like the commons, may be deemed 'appropriate' in the adjectival sense of being proper and suitable in particular circumstances, for particular needs, but commodification also recognises appropriation as a verb, the taking of something that rightfully belongs to others. In a world where use value and exchange value are themselves increasingly difficult to distinguish, a conception of the commons simply pitched as the antithesis of commodification may be hard to sustain. In some situations, for instance, resolutely *new* forms of the commons may be considered both symbiotic with and antagonistic to the cultures of commodification – as we see often enough in the digital domain. Thinking particularly of internetporn and its 'parasitism' on the libidinal and affective energies that feature so heavily in accounts of biopolitical production, Matteo Pasquinelli (2008: 208, emphasis in original) notes that: 'Accumulation still runs despite, or possibly thanks to digital *commonism*.' There is, at the very least, nothing *inherently* progressive in the political use of the commons, and just as the phrase and the practices it names can be marshalled in a manner inconsistent with the political Left, so too may it be hijacked by pro-capitalist regimes that allow further and further exploitation. As Silvia Federici (2010: 285) puts it, 'We must be very careful … not to craft the discourse on the commons in such a way as to allow a crisis-ridden capitalist class to revive itself, posturing, for instance, as the environmental guardian of the planet.' To take another pertinent example: Michael Goldman (1998), among others, has pointed in his critique of development discourse to the ways in which the commons metaphor has been mangled and misused by a generation of 'commons professionals', all in the service of 'domination and imperialism in North–South relations' (Goldman 1998: 22), thus restructuring the commons – which is to say, privatising them, at the behest of a crisis-ridden global capitalism (see also Radcliffe's critique in Chapter 8).

In the most cynical mood, we might speculate that a rapacious capitalism laying waste to the commons does not worry at all if the latter survives in some guises (for example, as public goods and pooled resources) to pick up the pieces, to shore up its operations and all of its consequences. The 'common' reduced to the 'public' becomes wholly complicit in neoliberalism's 'raiding and sacking of public finance' (Robinson 2014: 4): the public sector merely picks up the bill for the ravages of private wealth and power. It is all too easy for the capitalist order to socialise poverty and for the rest of us to give up on the *capital* that we already have lost to privatisation (Linebaugh 2008: 279). Here, the residual commons of public ownership – an underfunded, unsustainable public sector – is left to address the so-called negative externalities that the wealthy are able to avoid or bypass. We are left with two types of commons then: one soaking up the evicted and the expelled so that they pose no threat to capital, and the other collectivist but marginal and subordinated. The two are not interchangeable, and each presents a different form of politics: the former, a politics of public provisions and protections that may well meet popular needs, but does not inaugurate a different way of being in the world in the way that the latter does, but struggles to gain traction in an adverse institutional environment.

Perhaps the two ideas and their respective politics nevertheless need connecting, if only to lend momentum to a politics of commoning, whose transformative intentions find little sympathy in a mainstream in thrall to capitalism, and which are often jealously guarded by its advocates for the few and not the many, ending up horribly distorted by a populist politics confronted by all the inherent problems of mobilising 'the people' (Laclau 2006).[5] One such bridge, building on the history of institutions such as the Marshall Plan, the United Nations, the Welfare State, Universal Rights, Internationalism, Fair Trade, even the ideals and reality of National Parks and Garden Cities, is the recovery of an institutional field supposed to serve the general interest so that solidarity does not depend on social praxis alone. The commons returns here as a rule and possession – a necessary frame for commoning experiments – involving laws and policies blocking offshoring, irresponsible lending, monopoly, unregulated trade, and the privatisation of everything, and extending co-ownership, the right to be not excluded from property, the creative commons, and urban and natural resources, along with a host of reforms making for collective ownership, ethical food, banking, and trade, and the protection of vulnerable humans and non-humans. As several contributors intimate, without such a frame, however compromised in its historical actualisation, nothing at all stands in the way of the logic of encroachment and expulsion that has become institutionalised, nothing provides traction for the society of shared and protected resources. Here, the very narrative of the commons, with all its mixed meanings and ambiguities,[6] becomes productive, its words reason enough for not blaming the victims for its 'tragedy', for lamenting its loss, for valuing the many varieties of togetherness and things held in reserve, for 'bringing economic, social, and ethical concerns into greater alignment' (Bollier 2007: 29).

Words and meanings

But we are equally aware, and wary, that such crossings and reconciliations are far from straightforward, for it is not as though the spaces and practices of hegemons and communards are pure or hermetically sealed. We need to be careful about scripting normative conclusions into the politics of the commons and the 'manifold particularities' of commoning (Linebaugh 2008: 19). Too often, arguably, such a politics is valorised through a naive faith in 'democratic' proceduralism, supplemented by a proleptic invocation of new technologies and emergent formations, all underwritten by a normative fiat that judges, from the outside, whether a policy or a practice is 'progressive' or otherwise. Like Supreme Court Justice Stewart, and his famous test of the obscene ('I know it when I see it'), we all too quickly and comfortably assume that we will be able to pronounce upon the political value of commoning. But this hardly does justice to the variety of practices involved in making the common environment something to be spoken of and for, either in actual circumstances or in an idealised prospect. So, while Natalie Fenton and others rightly suggest here that new technology is particularly well suited for the allowance of emergent, provisional, even capricious solidarities and empathies, Adam Reed argues by cautionary contrast that the digital commons may not be perceived nor genuinely practised as a commons, but in fact quite the opposite – celebrating not mourning the passing of the commons. In other words, not every networked, horizontal social forum serves the interests of commoners, for its occupancy yields an associative field of diverse embodiments, interests and intentions, one lively enough to yield exclusionary outcomes, paradoxical as it might seem at first glance.

Even those who most insistently forward the sovereignty of the multitude acknowledge how easily the commons may be enclosed, deformed, diverted, or 'corrupted' (Hardt and Negri 2009). We might think here of cultural commons that promote distinction and serve to entrench class privilege, in the form of social and cultural 'capital', or alternatively those that sustain and police cultures of poverty. 'Collaboration' may be considered in its progressive sense of collective labour, or alternatively in its more sinister definition (see Schneider 2006). 'The common' is moreover so diverse that it resists easy incorporation into political platitudes or programmes. In addition to the example of the digital world, we may see the common incorporated in and sustaining of such complex phenomena as the informal or the black economy, organised crime and terrorism (however ill-defined and politically problematic these terms are), and a whole range of insurgencies and *émeutes*. In its 'corrupted' forms, the commons may extend even to the politics of the secret world of finance capital and its evasion of national taxation systems, at which point some may feel we have reached an aporia of sorts, given that financial offshoring might be both an attack on the commons and an instance of its corruption. But the larger point is that, if the commons is necessarily installed in all that it means to be human beings living in a shared social and natural environment, it is capable of being both produced as well as predated by capital. There is no necessary normative definition – the commons are always a kind of surplus or excess that refuses easy valorisation. Hardt and Negri (2006: 196) argue – simply, rightly, inadequately – that we have to learn to love some of the 'monsters' called up by the multitude and its 'common productive flesh' and to combat others. There is no further discussion of the so-called 'dark side' of the multitude (Bove and Empson 2002; Schneider 2006), nor the (by contrast, normatively neutral) 'parasitical' nature of the digital commons. As Pasquinelli puts it, 'Rather than the progressive destiny of networks, commons and collective knowledge, an ambivalent and unruly *animal nature* is positioned as the kernel of the political discourse – the ground traditionally removed from both liberal and so-called radical thought' (Pasquinelli 2008: 31). There are, clearly, and indeed entirely appropriately, multiple and multitudinous commons: such that distinguishing between positive and negative, beneficial and detrimental, creative and stultifying, between cooperation rather than collaboration, commons in the service of 'biopolitics' rather than in that of 'biopower', becomes the political point in question. It is not whether we are *for* rather than *against* the common, but precisely which forms of commoning to endorse, support and forward.

So while the language of the commons is a vital resource, it is hardly an unproblematic one. As Nicholas Blomley asserts in Chapter 6, although the response of an 'angry, moral, affective and political commons' may be normatively appealing, it is all too often largely rhetorical and typically undertheorised and vague. If in one view this imprecision is a weakness (and it is Gibson-Graham, Cameron, and Healy's great virtue to be, by contrast, so analytically secure in their contribution to this book), in another it could be seen as a kind of strength. Many commoners instinctively resist the attempt to arrive at a single definition on the commons. And, for all his reservations, Blomley observes that part of the political art of commoning is precisely to preserve its flexibility and innovation, avoiding the temptation of fixing the verbal noun. Peter Linebaugh has also argued that it is neoliberalism that has specialised in turning nouns into verbs, and then transforming them back again into 'hideous mouthfuls', 'as if to assert that the action and the struggle signified by the verb was over' (Linebaugh 2014: 6).[7] Perhaps thinking like a commoner means committing to the verb form in the attendant politics – making something of the creative potentiality of a space consisting of the many in agonistic relation, thinking of language dialogically rather than in the terms of a structural linguistics, preserving the sense that its practice can never *fully* be captured and commodified. What is needed seems to be a way of speaking about the commons that recognises its precious vulnerability while also doing justice to this creative, dynamic potential of commoning, and its 'manifold particularities' (Linebaugh 2008: 19). In this, a political narrative of the commons faces a signal dilemma, for sure, because if narratives are always examples of passionate thought, commons narratives carry the greatest political power when mobilised in terms of encroachment and usurpation (such as in Peter Linebaugh's ringing alarum: 'Stop, Thief!'). Are we able to recognise the uncommodifiable power of 'the common', and the irrepressibly creative practices of commoning, without taking something away from the necessary outrage that gives rise to counter-narratives and counter-politics? In the end, going forward might mean simply confronting the inevitable compromises. There is no avoiding of the thorny problem of what a politics in common, as Natalie Fenton has it, might actually look like – including the vaunted, emerging forms of collective politics, the new ways of imagining who 'we' are, new ways of bringing publics into being, new forms of co-production, and how we might mutually recognise others who are profoundly other, including the non-human other.

We indeed have to ask at this point: can there be a truly *cosmopolitan* politics of the commons that can serve these widely divergent interests, including the more-than-human, stretching across generations, and caring for the generations to come, to the future of the planet itself? With regard to the non-human, we might think here of the famous quatrain from the seventeenth-century folk poem often cited by commons advocates:

> The law locks up the man or woman
> Who steals the goose from off the common,
> But lets the greater villain loose
> Who steals the common from off the goose.

But while Linebaugh spots the hidden premise, or *enthymeme*, that stealing the common is *worse* than stealing the goose, he does not spot, or at least does not wish to mention, the *enthymeme* that equates stealing the goose with the goose's occupation of the common. Those whose motivation is for more-than-human interests will necessarily be uncomfortable with this rhetoric. Moving beyond such an anthropocentric understanding of the common (Winstanley's 'common treasury' of 1649 is very explicitly, and of course hardly surprisingly, a common treasury for 'man') is surely a priority if we are to do any justice to the demands of our 'common' environmental crisis. Practices of commoning need to be extended to a more-than-human community as well as to a more-than-capitalist one. These questions are among the most difficult of all when considering the shape of the present and future politics of commoning. Critics of the 'Anthropocene' have for instance rightly pointed out that in the larger scheme of things, there is no threat to the 'natural' global commons, only to the way 'we' live – though of course while the way we live (or die) is a matter of indifference to the planet on one level, species and environments continue to be adversely affected (Latour 2013; Vince 2014). It is tempting thus to pose the commons as the exposure to common *being*, and our creaturely life, not simply the practice of 'owning' public things. We may define the common as 'the sum of the pleasures, desires, capacities, and needs we all share' (Hardt and Negri 2006: 149), as well as the 'common human vulnerability' that emerges with life itself (Butler 2004: 31), and the recognition of common bodily existence and a 'metabolic commons' is indeed vital, as we come to reject the categorical separation of *bios* from *zoe* as a basis for understanding the 'common'. But commoners must be pressed to look past the *human* body as the 'biopolitical fabric of being' in mounting the case for its liberation, towards imagining how the shared biological existence of *all* life can be translated into the 'fabric of common experience' (Hardt and Negri 2009: 31, 124). The notion of the commons as collective life, a 'world-in-common' is intuitively more appealing: '*that dynamic mesh of relationships among creatures and their ambient surround that provide durable and livable architectonics for creaturely action and environmental sustenance*' (Reid and Taylor 2010: 8, emphasis in original). However necessary this translation, it will not be easy for humans to think and act *geologically*, as Kathryn Yusoff (2013) notes in her philosophical reflections on the ontologies of the Anthropocene.

Oneness and singularity

These complications and extensions of the keyword of this collection trouble the taken for granted in a further sense. Is it reasonable to say, as David Bollier (2007) has done, that: 'We, as citizens, own these commons'? What is this 'commonwealth' or 'common treasury'? How can we be sure that as citizens we can agree on how to protect and curate that which is 'ours'? All these qualifications seem problematically to restrict and ironically enclose what the commons is, and indeed what it might be. They raise doubts about the possibility of collective interest and common humanity (and its well-known imagery of lifeboats and spaceship Earths). Sarah Radcliffe here rightly bemoans the burgeoning 'fiction of a development commons' and its spurious investment in this supposedly 'common' humanity. Commoning, *com-unus*, 'coming together as one', does not have to mean a recognition of a common, pre-given identity, a collective bond; and common rights should never be confused with human rights (Linebaugh 2008: 44). This is at the very least for the fact that, as Karen Bakker has argued with water rights specifically in mind, 'Human rights are individualistic, anthropocentric, state-centric, and compatible with private sector provision of water supply; and as such, a limited strategy for those seeking to refute water privatization' (Bakker 2007: 447).

If the oneness of right-bearing individuals or citizen-owners turns out to be an improbable place holder for the commons, might a way forward be to relinquish designations of *having*, for those of *holding* in common? This shift of emphasis gestures towards an active political relationship, rather than some assumed common inheritance, natural or historical, a performance, a claim to territory, and a way of staking out the future as well as the past. Its task – and achievement – would be to place the politics of having and that of holding on an even footing. This would allow a politics of the commons to be freed from a nostalgia for community as the antidote of neoliberal individualism, with its 'simple appeal to homogenising or conservative ideas of collectivity' (Gilbert 2014: 161). It seems particularly obvious that we have to find ways of narrating the commons without lapsing into this language of community. Becoming one does not have to mean the dissolution of singularities into the single community or 'people'. Again, Karen Bakker has noted the limitations of the language of 'community' for the establishment of an alternative to the privatisation of water:

'the call for community control can sometimes fall into the trap of romanticizing communities, thereby denying the progressive potential of state–led redistributive strategies' (Bakker 2008: 247). Perhaps it is better to look to the alternative etymology of community, *com-munis*, and to argue that we are united not by what we *hold* in common but perhaps rather what we *owe*, our obligations, as the political philosopher Roberto Esposito (2009) suggests, to re-enfranchise people 'as responsible co-participants in the governance of the larger habitats that sustain them' (Menzies 2014: 184). With community recast as an obligation (*com-munis*) rather than a shared identity or interest (*com-unus*), commoning opens out to the world at large:

> This is how we must understand the slogan 'no commons without community'. But 'community' has to be intended not as a gated reality, a grouping of people joined by exclusive interests separating them from others, as with communities formed on the basis of religion or ethnicity, but rather as a quality of relations, a principle of cooperation and of responsibility to each other and to the earth, the forests, the seas, the animals.
>
> (Federici 2010: 289)

These are steps towards a politics of cooperation among those who have nothing in common, as Alphonso Lingis (1994) has expressed it. In the situation of what William Connolly (2010) calls the ever hastening 'minoritisation' of the world – the multiplication of all kinds of 'minorities' in the same territorial space – and we might add the dangers of decentralisation and local autonomy – the development of such a commoning that refuses the siren call of belonging, identity, or ownership, is as urgent as ever. Jeremy Gilbert puts this well, and succinctly:

> What is particularly useful about the idea of the commons as distinct from the idea of community is that it does not depend upon any presumption that the participants in a commons will be bound together by a shared identity or a homogeneous culture. Rather, they will be related primarily by their shared interest in defending or producing a set of common resources, and this shared interest is likely to be the basis for an egalitarian and potentially democratic set of social relationships.
>
> (Gilbert 2014: 165)

If advocates of the commons tacitly or explicitly accept this difference between shared identity and shared resource or interest, it can fairly be said that identity (as a derivative of the language and the ideology and the politics of property) is not fully dispensed with. This would require a conception of commons without community, without territory, perhaps without humanity itself; a conception of the commons that is normatively neutral and even innately suspicious of attempts to endow it with singular political meaning; a conception of the common that recognises that there isn't even a common concept of the common. We have thus to explore the language of singularities, interests, and sutures, not that of the communal or identitarian. We have to think the language of holding in common, not possessing. We have to avoid posing the commons in the terms of yet another 'Leviathan logic' (Gilbert 2014), in which, say, we all have an equal and identical stake as human beings, as persons.

What kind of politics might actually follow from these reservations? It may be that the best we can hope for is to proceed strategically, via diverse forms of intellectual and political enclosure, however much we are cautious about such moves. In this war of position, not all forms of enclosure are to be contested: as with opposition to capitalism, the enclosing gesture of the commons is at least a viable political strategy. It is always provisional, however, and the arts and practice of commoning have to come to terms with this necessary instability: for what is common is not a quality shared (such as 'common values', the 'common good', 'common sense', even a 'common [genetic] pool') so much as something produced through various kinds of cooperative and ethical relationship, including the affective, the discursive, and the political, between various kinds of necessarily differentiated actors or agents (and not just 'social' ones at that). Whatever 'common ground' there is, it cannot be fixed or settled, but is always-emerging, though care, collaboration, and altruism remain the fixtures. Marilyn Strathern's example in this book of organ transplantation strikes one cautiously but distinctly optimistic note here – for strangers may be nudged to become *collaborators*, with the installation of an opt-out mechanism – as opposed to the opt-in that simply threatens, in the science of xenotransplantation, more exploitation of non-human animals, because, in the austerity economics of altruism, 'we' as human beings cannot make good on the presumptions of a 'common humanity'. In any event, this hopeful practice gestures towards the promissory politics of cooperation rather than the assertive and illusory politics of solidarity. This kind of commoning involves not only what is, but also what might and should be: commoning is always about prospecting a commons-that-might-be as well as protecting the commons-that-are. It is about memory, but also imagination, and hope. It is about acting in certain ways and living ethically. The practical challenge for the future lies in finding ways of harnessing these open-ended expansions of 'coming together as one', however provisionally, and the fixities of things held in public, so that the practices of becoming and minoritisation are accompanied by large-scale and sustained inclusions.

Most immediately, the commons and commoning enjoins a certain academic and political practice. If, in Deleuzian terms, the 'collective' is a form of 'delirium', speaking *with*, writing *with* (Buchanan and Marks 2000: 4), we must learn to speak and to write in common, as the contributors to this symposium have done.

Conclusion

To think of the future of the commons, or perhaps better, of the futures of the common, is of course to name policies that can reinforce collective action and cooperative practice as a basis for human and ecological well-being, and across the landscape of operability, from land to ideas, resources to technologies. The success of attempts to secure a sustainable future depends on a clear politics of managing the cultural, economic, and environmental commons, committed to particular forms of organisation of gathered assets, juxtaposed diversities, and shared resources. This challenge at the interface of management and politics requires states, polities, legal entities, and international organisations to fashion techniques of government that can work effectively across a range of domains: from the political economy of markets and nations, to the rights of peoples, citizens and distant others, and the configurations of infrastructures, ecosystems, and public services and spaces. The many practical interventions suggested by the contributors in this book – an environmental 'paracommons', the 'offshore' brought back under national democratic control, the right not to be excluded from property, a strengthened international rights framework, universal basic income without moral conditions, commoning communities, social technologies for urban metabolic equity, and infrastructures of intimate publics – all attest to the necessity and possibility of this way of recovering the commons, as of course does the expanding literature, following the lead of Elinor Ostrom (2012), on who should govern the commons.

But in this opening chapter, we have tried to show that how we imagine the commons is equally important, for the keyword (we accept, the buzzword) comes with many mystifications, meanings, and compromises, all of which have a bearing on what should be done, and by whom, and in whose name. In making the commons an object still to be defined, our aim is to open up and pluralise possibility, and to ask all the time what a *politics* of possibility entails: that is, its challenges, constituents, and modalities. This will be a politics of jurisdiction, rights, and rules – new institutional fixes – but it is also one, as amply shown by the chapters, of preparing the ground for a shared and commonly curated world. Amid the ever-present politics of expulsion and appropriation persist practices of commoning, strategies of cooperation, agonistic occupancies; practices that form the ground of living collectively and with care for human and non-human others, and through which the political skills of organisation, opposition, and negotiation are honed (Mitchell 2012; Vasudevan 2015). These practices are fragile and dispersed, no doubt, but they are plentiful and they are formative. The powers that benefit from enclosure and appropriation will not cede unless forced to, and in the absence of any organised Left surge in or out of government for radical change (though Syriza, Podemos, and some Latin American democracies are notable exceptions), the learning and leaning acquired in the micro-worlds of commoning may turn out to be pivotal. The lived experience of acting cooperatively and ethically reverberates politically, and in the overlaps of the intimate publics binding strangers, it amplifies and lingers, the stay and the strength of all those arts that push back at the encroachments of privation and self-interest. These are political arts, let us be clear, arts that weave the experience of being in common with the tactics learnt of defending that way of being (Amin and Thrift 2013), creating a breathed ambient atmosphere (McCullough 2013) that makes any other 'air', including the well-rehearsed theme of a commons wrecked, *feel* politically inadequate as well as morally illegitimate.

Even thinking the commons in this way – as an *enrolment* – can be considered a form of affective political practice (practice in the sense that all arts can be practised), as a way of generating 'new spaces of the common', as Esposito (2013) has it. All forms of associational practices such as this aim to make the empty rituals of corporatism seem odd, the necessity of another way of being in the world a demand. We note and support these new and emergent commons, and in this sense reject the all-too-familiar theme of shrinking and crowding out. This will be the ground ultimately for suspicion of any relict, mercantilist sense of the commons – as if the logic of scarcity governed the nature of all of the common resources upon which we can draw for the sake of our collective future – and of the commons articulated only in the preterite tense, as the world we have already lost. But it is also the ground of thinking past easy but misleadingly straightforward dichotomies, between the commons and enclosure, for instance, between private self-interest and public-facing altruism, between private right and common right, or even perhaps between the commons and the world of property, for there are a variety of forms of ownership and non-ownership – and 'property' itself is a rather more ambiguous term than Hardt and Negri's conception of the 'republic of property' allows (see Giordano 2003).[8]

Releasing the commons, then, is surely as much an art as it is an act, unleashing a host of campaigns in its name and under its banner, and relishing the contagion of collaborative work in shared space, but it is a practice of imagination too, allowing many vocabularies of being in common to proliferate. One shared achievement would be to make kin-making with diverse humans and non-humans natural (Haraway 2015), to render thinking like a commoner (Bollier 2007),

> no goodwill dream of going back to an innocent idealized past [but] a pragmatic challenge, entailing no fairy tales, no wishful thinking but an ongoing care and concern for the fragility of the assemblage, for the maintenance of what is always a more than human interdependence.
>
> (Stengers 2014: 7)

The new 'interagentivity' *between* humans and *with* nature, the 'multiplication of diplomatic scenes' of co-habitation and care-full negotiation, will better prepare us for the collective task, as Latour (2014) puts it, of 'composing' the common world.

Notes

1 See for instance, Menzies 2014, or Linebaugh 2008.
2 Just as Rebecca Solnit (2014: 4) reflects that 'This territory to which I am, officially, consigned couldn't be more spacious, and I couldn't be more pleased to be free to roam its expanses,' so we see the advantage of this book's breadth and range, and also the appropriate stance towards commoning.
3 While we share much of Hardt and Negri's ambivalence towards the reduction of the common or the commons to the public, like them we see no *necessary* incompatibility between reformist and radical agendas.

4 *Commodity* was established in English in the fifteenth century, deriving from the French *commodité*, involving benefit or profit. The French word itself stems from the Latin *commoditatem* (nominative *commoditas*), meaning 'fitness' or 'adaptation'. The root (which has resulted in the English terms *commodious* and *accommodate*, among others) invokes 'appropriate' and 'proper' as well as advantage and benefit.

5 Note that constructing 'the common sense of the common people' (Patten 1996) is quite easily hijackable by the political Right, in the imagination of a closed, homogeneous society, as we see from contemporary populist mobilisation against globalisation, cosmopolitanism, and austerity.

6 For example, if the 'commons' implies moving from the Me to the We, who and what is the 'we'? Are 'we' speaking for 'humanity' (acting 'as a species'), and/or the planet? Are 'we' speaking for the many – the 99 per cent say – as opposed to the few, the common enemy of the new global aristocracy (Robinson 2014)? Or are we to invoke at the same time the 'multitude' and '*the* poor' – simultaneously plural and singular – as Hardt and Negri seem to prefer?

7 But even so, in his earlier work, he can see that the notion of commoning, the verbal form, is merely another potential trap (Linebaugh 2008: 279).

8 David Harvey (2011) has complained that certain kinds of questions are avoided by positing the tragedy of the commons as one of common ownership, rather than of private property, but, again, different kinds of property may be conceived.

References

Amin, Ash and Nigel Thrift. *Arts of the Political: New Openings for the Left*. Durham, NC: Duke University Press, 2013.

Anderson, Alun. *After the Ice: Life, Death, and Geopolitics in the New Arctic*. Washington, DC: Smithsonian Books, 2009.

Bakker, Karen. 'The "Commons" Versus the "Commodity": Alter-Globalization, Anti-Privatization and the Human Right to Water in the Global South'. *Antipode* 39, no. 3 (2007): 430–55.

Bakker, Karen. 'The Ambiguity of Community: Debating Alternatives to Private-Sector Provision of Urban Water Supply'. *Water Alternatives* 1, no. 2 (2008): 236–52.

Bollier, David. 'The Growth of the Commons Paradigm', in Charlotte Hesse and Elinor Ostrom (eds), *Understanding Knowledge as a Commons*. Cambridge, MA: MIT Press, 2007, 27–40.

Bove, Adrianna and Erik Empson. 'The Dark Side of the Multitude'. *Dark Markets: Infopolitics, Electronic Media and Democracy in Times of Crisis, International Conference, Vienna*. 2002.

Buchanan, Ian and John Marks. *Deleuze and Literature*. Edinburgh: Edinburgh University Press, 2000.

Butler, Judith. *Precarious Life: The Powers of Mourning and Violence*. London, Verso, 2004.

Connolly, William E. *A World of Becoming*. Durham, NC: Duke University Press, 2010.

Eizenberg, Efrat. 'Actually Existing Commons: Three Moments of Space of Community Gardens in New York City'. *Antipode* 44, no. 3 (2012): 764–82.

Esposito, Roberto. *Communitas: The Origin and Destiny of Community*. Stanford, CA: Stanford University Press, 2009.

Esposito, Roberto. 'Community, Immunity, Biopolitics'. *Angelaki: Journal of the Theoretical Humanities* 18, no. 3 (2013): 83–90.

Federici, Silvia. 'Feminism and the Politics of the Commons in an Era of Primitive Accumulation'. *Uses of a Whirlwind: Movement, Movements, and Contemporary Radical Currents in the United States*, edited by Craig Hughes, Stevie Peace, and Kevin Van Meter for the Team Colors Collective. Oakland, CA: AK Press, 2010, 283–93. Also available at: www.commoner.org.uk/?p=113 (accessed 18 January 2016).

Gibson-Graham, J.K. *The End of Capitalism (As We Knew It): A Feminist Critique of Political Economy*. Oxford: Blackwell, 1996.

Gilbert, Jeremy. *Common Ground: Democracy and Collectivity in an Age of Individualism*. London: Pluto, 2014.

Giordano, Mark. 'The Geography of the Commons: The Role of Scale and Space'. *Annals of the Association of American Geographers* 93, no. 2 (2003): 365–75.

Goldman, Michael. 'Introduction: the Political Resurgence of the Commons', in Michael Goldman (ed.), *Privatizing Nature: Political Struggles for the Global Commons*. London: Pluto Press, 1998, 1–53.

Haraway, Donna. 'Anthropocene, Capitalocene, Plantationocene, Chthulucene: Making Kin'. *Environmental Humanities* 6 (2015): 159–65.

Hardt, Michael and Antonio Negri. *Multitude*. New York: Penguin, 2006.

Hardt, Michael and Antonio Negri. *Commonwealth*. Cambridge, MA: Harvard University Press, 2009.

Harvey, David. 'The Future of the Commons'. *Radical History Review* 109 (2011): 101–7.

Laclau, Ernesto. 'Why Constructing a People is the Main Task of Radical Politics'. *Critical Inquiry* 32, no. 4 (2006): 646–80.

Latour, Bruno. *An Inquiry into Modes of Existence: An Anthropology of the Moderns*. Translated by Catherine Porter. Cambridge, MA: Harvard University Press, 2013.

Latour, Bruno. 'Another Way to Compose the Common World'. *HAU: Journal of Ethnographic Theory* 4, no. 1 (2014): 301–7.

Linebaugh, Peter. *The Magna Carta Manifesto: Liberties and Commons for All*. Berkeley, CA: University of California Press, 2008.

Linebaugh, Peter. *Stop, Thief! The Commons, Enclosures, and Resistance*. Oakland, CA: PM Press, 2014.

Lingis, Alphonso. *The Community of Those Who Have Nothing in Common*. Bloomington, IN: Indiana University Press, 1994.

McCullough, Malcolm. *Ambient Commons: Attention in an Age of Embodied Information*, Cambridge, MA: MIT Press, 2013.

Menzies, Heather. *Reclaiming the Commons for the Common Good*. Gabriola Island, BC: New Society Publishers, 2014.

Mitchell, William, J.T. 'Image, Space, Revolution: the Arts of Occupation'. *Critical Inquiry* 39, no. 1 (2012): 8–32.

Ostrom, Elinor (with contributions from Christina Chang, Mark Pennington, and Vlad Tarko). *The Future of the Commons: Beyond Market Failure and Government Regulation*. London: Institute of Economic Affairs, 2012.

Pasquinelli, Matteo. *Animal Spirits: A Bestiary of the Commons*, Rotterdam: NAi Publishers, 2008.

Patten, Steve. 'Preston Manning's Populism: Constructing the Common Sense of the Common People'. *Studies in Political Economy* 50 (1996): 95–132.

Reid, Herbert G. and Betsy Taylor. *Recovering the Commons: Democracy, Place, and Global Justice*. Chicago, IL: University of Illinois Press, 2010.

Robinson, William I. *Global Capitalism and the Crisis of Humanity*. Cambridge: Cambridge University Press, 2014.

Sassen, Saskia. *Expulsions: Brutality and Complexity in the World Economy*. Princeton, NJ: Princeton University Press, 2014.

Schneider, Florian. 'Collaboration: the Dark Side of the Multitude'. *Sarai Reader 2006: Turbulence* (2006): 572–6.

Solnit, Rebecca. *Wanderlust: A History of Walking*. New York: Penguin, 2001.

Solnit, Rebecca. *A Field Guide to Getting Lost*. New York: Penguin, 2006.

Solnit, Rebecca. *The Encyclopedia of Trouble and Spaciousness*. San Antonio, TX: Trinity University Press, 2014.

Stengers, Isabelle. 'Gaia, the Urgency to Think (and Feel)'. Paper presented at 'The Thousand Names of Gaia' conference, Rio de Janeiro, September 2014, at: https://osmilnomesdegaia.files.wordpress.com/2014/11/isabelle-stengers.pdf (accessed 18 January 2016).

Vasudevan, Alex. 'The Autonomous City: Towards a Critical Geography of Occupation'. *Progress in Human Geography* 39, no. 3 (2015): 316–37.

Vince, Gaia. *Adventures in the Anthropocene*. London: Chatto and Windus, 2014.

Williams, Colin C. *A Commodified World?: Mapping the Limits of Capitalism*. London: Zed Books, 2005.

Yusoff, Kathryn. 'Geologic Life: Prehistory, Climate, Futures in the Anthropocene'. *Environment and Planning D: Society and Space* 31, no. 5 (2013): 779–95.

2 The commons and offshore worlds

John Urry

The commons

The fate of what is held in common has never appeared as problematic as today. Climate change, population growth, extreme weather, food and water shortages, and competition for energy threaten the global commons of planet earth, as well as many other smaller-scale commons. Much that has been thought of as 'public' or 'common' has been in various ways privatised, enclosed, or commercialised. Even research and knowledge relating to the commons is often now commercially confidential and less subject to public visibility and scrutiny.

This privatising of what had been common has not occurred without material struggles and imaginaries of collective ownership, well-being, and use. There have been various material, discursive, and ideological processes through which different entities have been held, managed, and imagined as in some manner common. This chapter examines under-explored elements of this terrain of struggle around the commons and the public. The commons are a terrain of struggle but it is hard to be optimistic as to the outcome of these fateful struggles given the growth of what I will analyse as 'offshore worlds'. I use the term commons here to refer to a wide array of collective and/or public discourses and processes that are jeopardised by the proliferation of offshore worlds.

The 1980s onwards saw the striking emergence of various interdependent global processes. Many believed that the global movements of money, people, ideas, images, information, and objects were broadly beneficial. Most aspects of contemporary societies were thought to be positively transformed through increased borderlessness and the forming of many new kinds of commons, and especially virtual commons. Writing in 1990, Ohmae described this emergent borderless world:

> the free flow of ideas, individuals, investments and industries ... the emergence of the interlinked economy brings with it an erosion of national sovereignty as the power of information directly touches local communities; academic, professional, and social institutions; corporations; and individuals.
>
> (Ohmae 1990: 269)

This optimism during what Stiglitz termed the 'roaring nineties' was thought to be engendering novel businesses, cosmopolitan polities, international friendship, family lives, international understanding, and a greater openness of information and communications (Stiglitz 2004). Many believed that societies were invigorated through these flows of ideas, information, and people, making societies more 'cosmopolitan'. Borderlessness seemed to sustain a new sense of a shared planet, with the first report of the Intergovernmental Panel on Climate Change appearing in 1990. Especially significant was the utopic 'invention' of the web by Tim Berners Lee around 1990 with the resulting proliferation of multiple virtual worlds which transformed economic and social life. The development of various 'virtual commons' contributed to a 1990s 'global optimism' as to a progressive open 'common' future (Urry 2014: Chapter 1).

But this 1990s decade did not turn out to be the harbinger of a long-term, optimistic, and borderless future. It transpired that there are many dark sides to all this movement; much is not at all 'common'. Migrating across borders are not just consumer goods and services and a new more open commons. Also moving over borders are environmental risks, trafficked women, drug runners, laundered money, terrorists, international criminals, outsourced work, climate change refugees, slave traders, asylum seekers, property speculators, smuggled workers, waste, financial risks, and untaxed income. These flows across borders made the achievement of common interests much more problematic as multiple borders and new fierce boundaries developed, multiplied, and were securitised, often using similar techniques to the once open and supposedly democratic web.

Especially significant here are the implications of 'offshore worlds' for holding, managing, and imagining 'in common'. In some cases offshore worlds have been strategically developed to facilitate the specific privatising of various commons. New physical, economic, and virtual borders or boundaries have been generated by what Warren Buffett, probably the world's most successful investor, describes as the 'rich class'. He maintained: 'There is class warfare, all right, but it's my class, the rich class, that's making war, and we're winning' (Farrell 2011; and see Sayer 2015). One element of that class war involved generating offshore worlds undermining various kinds of 'commons'. This chapter asks whether it is possible for ideas of the commons to co-exist with such extensive offshore worlds.

Secret worlds

Brittain-Catlin describes the effects of such offshoring: 'the negative, dark spirit … today pervades the offshore world and its network of secret paraphernalia and hidden practices that are so closely bound into the global economy' (2005: 118).[1] There has been a striking proliferation of various 'secret worlds' as globalisation generated not openness and multiple commons but their very antithesis, many 'offshore worlds' relying upon secrets and lies.

Such offshoring involves moving resources, practices, peoples, and monies from one national territory to another but hiding them along and within various secrecy jurisdictions. Rules, laws, taxes, regulations, and norms are evaded. It is all about rule-bending and rule-breaking, getting around rules in ways that are illegal, or go against the spirit of the law, or which use laws in one jurisdiction to undermine laws in another. This offshoring world is dynamic, reorganising economic, social, political, and material relations between societies and within them. Populations and states find that more and more resources, practices, peoples, and monies are made or kept secret and opaque. The global order is the opposite of the commons – it is one of concealment, of secret gardens principally orchestrated in and for the rich class (see details in Elliott and Urry 2010).

These worlds were made possible by new or globalising mobility systems. These include container-based cargo shipping; aeromobility with many new airport spaces; the internet and new virtual worlds; extensive car and lorry traffic; new electronic money transfer systems facilitating legal and illegal flows; taxation, legal, and financial expertise often oriented to avoiding national regulations; and 'mobile lives' involving frequent legal and illegal movement across borders. Especially significant within most of these offshored worlds are delocalised virtual environments enabling information, money, trades, images, connections, and objects to move digitally along routeways in the shadows, often in the deep or dark web not indexed by normal search engines. Virtual environments are part and parcel of contemporary offshoring and results in delocalising production, consumption, and sociability over the past few decades.

These processes of 'offshoring' range from those where there is a mere dependence upon overseas resources, to those which are onshore but which enjoy offshore status and may be concealed, to those which are literally out to sea, over the horizon, secret, and often illegal. Offshoring has developed into a generic principle of contemporary societies, and it is thus impossible to draw a clear divide between what is now onshore and what is offshore (see Brittain-Catlin 2005).

Three-quarters of the surface of the earth is water, water having often been presumed to be a key 'commons' resource. The seven billion humans are crowded onto just one-quarter of the earth's surface, the rest are oceans. These watery worlds provide ways to assemble as secret what would otherwise be onshore and visible (on watery worlds, see Anderson and Peters 2013). Almost all this vast ocean world is out of sight. Oceans contain many unregulated 'treasure islands' of low tax and much pleasure; ships fly flags of convenience with conditions of construction and work driven to the bottom; oceans are places where desperate poor migrants routinely lose their lives; oceans are a global rubbish dump with the Great Pacific Garbage Patch said to be twice the size of France; and there are unregulated climates as the outlaw sea subjects humans to its heightened unruliness, as intense storms, hurricanes, storm surges, rising sea levels, and flooding boomerang back onto land from the oceans (Hansen 2011; Urry 2014: Chapter 9). The sea might be described as a neoliberal paradise, a world almost without government, taxes, laws, and where only powerful ships and their companies survive, with the rest often literally sinking to the bottom. The sea is an unruly space not really owned and governed by states; it is mostly risky, free, and unregulated. Contemporary oceans demonstrate the tragedy of the commons as Langewiesche (2004) describes the 'outlaw sea'.

Offshoring documents many of these offshored worlds (Urry 2014). These include: offshoring manufacturing production to cheaper sites; systematically reducing tax liabilities and hence heightening inequalities; creating many secret offshore companies; engendering new forms of financialisation; developing expertise in novel ways of marginalising workforces through outsourcing work; forming new places of pleasure away from observation by friends/family; extracting infrastructural investment from states; externalising the costs of waste and emissions to other countries; using moments of crisis to force through neo-liberal restructuring of these processes; mobilising various discourses promoting marketisation; and creating astonishing new products that are based upon new 'needs' including for security. These offshore worlds all derive from the global freedom to move monies, income, wealth, people, waste, and loyalties from pillar to post. This dizzying 'mobile' world, bright and dark, open and secret, free and destructive, is formidably difficult to regulate in 'common' interests and that of course is often their point.

Moreover, these developments were no accident. They did not just occur but resulted from sustained efforts to reverse 'Keynesianism'. Across much of Europe and north America from the Great Depression of 1929 onwards, Keynes' thought had been central in showing how economic systems would not rectify unemployment and economic depression (Keynes 1936). Economies were not self-regulating and would not automatically restore equilibrium. He powerfully argued for the virtues of counter-cyclical tax-funded state expenditures, systems of national planning, and the idea that there was a collective national interest separate from the interests of specific individuals and companies. From the 1930s to the 1970s the dominant Keynesian discourse was that states had the capacity to rectify market deficiencies and should do so. Such Keynesianism involved support for many kinds of state intervention, for the idea that the national economy is a commons as opposed to the interests of specific corporations and rich individuals.

However, Keynesianism was soon challenged. In various secret societies and meetings during the long period of post-war Keynesianism the post-war capitalist class plotted offshore worlds. As early as 1947 a senior bank official brought together various scholars to a secret meeting at Mont Pèlerin, near Geneva, under the direction of Friedrich Hayek (see his *The Road to Serfdom*, 1944). This Mont Pèlerin Society funded by Swiss banks it should be noted was central in a global fightback against Keynesian support for state interventionism. This struggle to reverse Keynesianism was organised through many further secret meetings (see details in Stedman Jones 2012: Chapter 2). One key person at these many meetings was Milton Friedman, who was central in developing so-called 'neoliberalism' which from the late 1970s gained massive global traction. There has been a titanic struggle between two discourses about markets and states, between 'state-ism' and 'neoliberalism' with the latter 'winning' over the former during the past three to four decades (see the illuminating Klein 2007).

Offshoring finance

During the era of organised capitalism or Keynesianism in the 'West', lasting roughly from the 1940s until the 1970s, each national state controlled what its banks and other financial organisations were able to do, especially through the state setting firm and reasonably high reserve requirements (Lash and Urry 1987, 1994).

But as 'disorganised' or neoliberal capitalism was developing from the 1970s onwards, many such requirements disappeared and conditions were set for off-shoring income, wealth, and much else. As the firewalls disappeared so a deter-ritorialised system expanded and there was increasing competition between places to provide offshore accounts offering secrecy, lower tax rates, and less regulation (on the importance of firewalls, see Haldane and May 2011). Many new tax havens and offshoring became central to finance as it spun off from the national regulations core to organised capitalism. And, with the growth of global inequality, many individuals and corporations increasingly provided much income for developing societies to move into providing offshored financial services for the powerful rich class. Palan, Murphy, and Chavagneux show how tax havens are core to developing this world of finance; although each haven is not significant on its own: 'combined, they play a central role in the world economy ... one of the key pillars of ... "neoliberal globalization"' (2009: 236).

Thus what emerged within contemporary capitalism was a significantly untaxed, ungovernable, and out-of-control 'casino capitalism', more like gambling than banking and which has magnified economic, social, and property inequalities in most countries across the globe (Piketty 2014; Sayer 2015). Since the 1980s there has been an astonishing growth in the movement of finance and wealth to and through the world's sixty to seventy tax havens, representing more than one-quarter of contemporary societies. These havens include Switzerland, Jersey, Cyprus, Macao, Cayman Islands, British Virgin Islands, Madiera, Monaco, Panama, Dubai, Liechtenstein, Singapore, Hong Kong, Dubai, Gibraltar, the City of London, and Delaware (see Shaxson 2012). These 'secrecy jurisdictions' were core to neoliber-alising the world economy from 1980 once many exchange and related controls were removed from the late 1970s onwards. To be offshore is to be in paradise, by contrast with the high-state-high-tax life experienced onshore. Tax havens are places of escape and freedom, a paradise of low taxes, wealth management, deregulation, secrecy, and often nice beaches as well.

Offshoring involves getting around rules, being 'irresponsible'. Most offshoring practices are not incidental but systemically engineered and legally reinforced to avoid regulations, to be kept secret and to 'escape' offshore, helping to form and sustain multiple intersecting offshore worlds. Palan, Murphy, and Chavagneux emphasise that tax havens are deliberately created entities designed to 'smooth' the transactions of those not normally resident within a society's borders and whose transactions are cloaked in veils of secrecy (2009). Such tax havens or secrecy jurisdictions presuppose the active work of accountants, bankers, lawyers, and tax experts to establish governance designed as opaque and benefiting those who are not normally 'legal citizens' of that particular society.

Elites thus escape many kinds of formal and informal sanctions and set up conditions for further extending their income and wealth. This inaccessibility makes them less responsible for their actions, especially in those very societies in which their actions would appear to occur. They are literally irresponsible being able to 'escape', this having become a key feature of power according to Bauman within an offshored world (2000; on 'unrestrained greed', see also Tett 2010).

Money staying onshore is almost the exception, suitable only for 'little people' still paying tax. Most big money is offshored. Overall offshore has been generated by and favours large corporations. The offshore world makes it hard for 'innovative minnows' to compete and, if they do prosper, they will become parts of large financial or corporate bureaucracies whose income flows will be significantly offshored. The world of offshore systemically weakens local, smaller companies, which are confronted with nothing like a level playing field as they compete with large offshored companies and their many offshore accounts (Urry 2014: 1–2).

Almost all major companies possess offshore accounts/subsidiaries often running into the hundreds; more than half world trade passes through them; almost all high net worth individuals possess offshore accounts enabling tax 'planning'; and ninety-nine of Europe's 100 largest companies possess off-shore subsidiaries. It has been calculated that offshored money has grown from US$11 billion in 1968 to US$21 trillion in 2010 (now equivalent to about one-third of annual world income). Fewer than ten million people currently own this US$21 trillion offshore fortune, a sum equivalent to the combined GDPs of the USA and Japan. This is *the* source of power and wealth of the super-rich with almost all owing their fortunes to the rapid and secret moving of money and ownership 'offshore' (see estimates in Palan *et al.* 2009: Chapter 2; and Urry 2014: Chapter 4).

Overall Shaxson shows that offshore is how the world of power now works (2012: 7–8). Most big money is offshored although offshore includes most mainstream banks and financial institutions. Corporations and high net worth individuals within the rich North have come to depend upon these offshore financial centres often located in what were fairly poor developing societies and which were often assisted in their 'development' by the centres of financial power such as the City of London.

Worldwide there are thought to be twelve million high net worth individuals, each holding at least US$1 million in investable assets. The combined wealth of this 'rich class' is US$46 trillion, a sum equivalent to two-thirds of annual world GDP.[2] The annual loss of taxation from this offshoring world is hundreds of billions of US dollars.[3]

Nevertheless for all the significance of offshore centres in 'developing countries' such as Caribbean islands, Shaxson documents how the USA 'is the world's most important secrecy jurisdiction' (2012: 146). In the little state of Delaware there is a single building housing 217,000 companies, the largest and probably the most unethical building in the world. More than one million business entities are incorporated in Delaware, including over 50 per cent of all publicly-traded companies based in the USA.

Successful offshore tax havens are the opposite of a commons. A 'good' tax haven should possess some or ideally all of the following features: it should not tax income, profits, or inheritance; its banks should offer various currencies, operate online, and not require personal visits; new banks accounts should require minimal documentation; there should be bank secrecy with no Tax Information Exchange Agreements with other countries (about forty havens have no such agreements); and bank accounts can be opened using an 'anonymous bearer share corporation' with the consequence that people's names do not appear in any public registry or database. A tax haven seeks to deliver secrecy and discretion, to 'ask no questions, tell no lies'.

The banking sector is the most prolific user of tax havens, with over half of the overseas subsidiaries of major banks located in 'treasure islands' of low tax. The company called Goldman Sachs Structured Products (Asia) Limited is based in the tax haven of Hong Kong.[4] It is controlled by another company called Goldman Sachs (Asia) Finance, which is registered in another tax haven, namely Mauritius. That is administered by a further company in Hong Kong, which in turn is directed by a company located in New York. This is controlled by another company in Delaware, a major tax haven, and that company is administered by yet another company, also in Delaware, GS Holdings (Delaware) L.L.C. II. This in turn is a subsidiary of the only Goldman Company that most people have heard of, the Goldman Sachs Group that occupies a glitzy tower in Battery City Park in New York. In 2012 this company generated a worldwide turnover of around US$34 billion, employing nearly 30,000 staff. This particular chain of ownership is one of hundreds of such chains that exist within the single company Goldman Sachs.

These big companies are thus built rather like Russian dolls, with multiple layers of secrecy and concealment. In major research by ActionAid it was reported that advertising giant WPP held 618 offshore accounts, HSBC 496, Royal Dutch Shell 473, Barclays 471, BP 457, RBS 393, Lloyds 259, British Land 187, and Prudential 179.[5]

So generally there is much financial intermediation and this hollows out large industrial corporations with ownership increasingly vested in financial institutions concerned with short-term 'shareholder value'. Indeed ownership can on occasions last less than a second through algorithmic trading! The large industrial corporation became outdated, their numbers halving over the past twenty years (Davis 2009, 2012). Corporate cultures providing reasonably generous benefits for 'corporation men' are less common. Sennett and others lamented the resulting decline in people's long-term commitment and character to their workplaces that this new economy engenders (Sennett 1998). Much intermediation is directed by new interdependent financial elites that emerged over the past few decades (Savage and Williams 2008). This power of financial intermediation and speculation runs counter to the interests of an economy made up of smaller companies innovating and producing new products and services, and especially to those implicated in developing a lower carbon economy-and-society.

Centrally important here in the growing power of finance, circulation, and debt are private equity buyouts (Appelbaum and Batt 2014). These involve a private equity firm putting up a small proportion of the purchase cost of a company or part of a company. The rest comes from institutional investors or is borrowed using the future company's assets as collateral. Once the public company has been bought, it will be made 'private' and hidden out of sight. There are far fewer restrictions on what private entity firms are able to undertake partly because they are treated as investors rather than employers. Evidence shows that, through exemptions in securities laws, most private equity funds avoid regulatory oversight. And yet private equity-owned firms are more likely to reduce employment levels, have slower growth rates, and go bankrupt more often. Private equity-owned firms, being registered and operating offshore, are also more likely to deploy tax avoidance/evasion strategies and are much harder to sign up to the pursuit of 'common' objectives.

So we have seen that offshore worlds facilitate the growth and power of the 'rich class' that increasingly meet as secret societies, such as the offshored annual meetings of the Bilderberg Group. As elites circulated spatially so they developed connections to extend further offshored worlds and its discourses. Private meetings and more public think-tanks helped to orchestrate this offshore world and its corporate, individual, and policy world beneficiaries.

One commentator thus reports that for billionaires: 'you don't live anywhere, and neither does your money. Or rather you live everywhere, and so does your money' (quoted in Featherstone 2013: 115). This offshore world involves rapid movement across the oceans, homes dotted around the world, endless business travel, private schools, family life structured around occasional get-togethers, private leisure clubs, luxury ground transport, airport lounges, private jets, luxury destinations, and places of distinction and luxury for encountering other super-rich (on these multiple mobilities, see Birtchnell and Calterio 2013). Place, property, and power are intertwined so forming and sustaining such a networked and often hidden rich class able to avoid the commons and generating complex private routeways (on rapidly developing private jet services, see Budd 2013).

This growth of inequality engenders powerful interests to protect and further extend the bases of such unequally distributed global income and wealth. Offshoring is part and parcel of the realizing of such unequal interests. And such inequalities matter a great deal, since access to 'services' increasingly depends upon each person's income and wealth; the more unequal these are, the less chance there is that people will be regarded as in any way equal (Sayer 2015). The rampant marketisation of almost everything crowds out many other reasons why people may act towards each other, such as common notions of fairness, service, duty, and sociability.

The fight back

Interestingly though since around 2,000 issues of offshoring have become significantly contested. In particular tax evasion and aggressive tax avoidance by the rich and powerful (such as UK Prime Minister David Cameron's father) as well as the role of tax havens (in many crimes) have become central to an emerging fiscal counter-politics. From the turn of the millennium a new taxation politics emerged. There are many critical reports (by Oxfam and other charities on how tax havens contribute to global poverty); media stories (even in the pro-business *Wall Street Journal* or *The Economist*); new campaigning NGOs (such as Offshore Watch); interventions by the World Social Forum (establishing a global campaign against tax havens); new kinds of research capability (such as the Tax Justice Network); a greater role for the OECD and EU seeking to curb 'unfair tax competition'; an increased rate of leakage of financial data to the media (thirty-eight media organisations received such data in March 2013); and the raised public identification and critique of top corporations that we now know are 'scandalously' involved in aggressive tax avoidance/evasion. Tax moved from being a private to a public issue as issues of 'dodging' (which includes both evasion and aggressive tax avoidance) went mainstream. The media report tax stories with increasing regularity.

This taxation counter-politics, this assertion of the interests of the 'commons', is a torrent. No longer is tax a private matter just for oneself and one's accountant! Much direct action, NGO activities, official government reports, and a new activism have exposed and denigrated many different forms and aspects of 'tax dodging'. Such dodging is seen as reducing the capacity to tax revenues where income and wealth are generated and as undermining a level playing field, since local companies normally pay full taxes while transnational corporations do not.

Tax shaming is rapidly developing into a global movement. This was strikingly shown in early 2013 when the International Consortium of Investigative Journalists received in their mail a computer hard drive packed with corporate data and personal information and emails. This information totalled more than 260 gigabytes of data. It originated from ten offshore jurisdictions and included details of more than 122,000 offshore companies, nearly 12,000 intermediaries (agents or 'introducers'), and about 130,000 records relating to those who run, own, benefit from, or hide behind offshore companies. It showed that those setting up offshore entities most often lived in China, Hong Kong, Russia, and former Soviet republics. Many positions are held by so-called nominee directors whose names can appear in hundreds of companies. Nominee directors are people who, for a fee, lend their names as office holders of companies that they know little about and have certainly not examined for their financial and ethical probity.

Other authors have begun to generalise this issue of tax, some claiming, for example, that 'tax is a feminist issue' (Walby 2009). Large corporations and rich individuals are increasingly forced to defend their tax position, often seeking a 'taxwash' to keep the scandal-hungry media and protestors at bay.

This idea of taxation as a kind of commons increasingly threatens the world's major brands, critiqued for their aggressive tax avoidance and deliberate evading of transparency and public scrutiny. The widespread development of 'tax shaming' reveals the power of the commons to fight back. And other aspects of offshoring such as waste and especially e-waste, CO_2 emissions outsourced to China, and tortured bodies stemming from extraordinary rendition have also been subject to increasing protest within the media and social media, and especially on the streets. These offshore worlds are increasingly contested by strategies of 'reshoring' the commons.

Regulating offshore?

Offshore, we might say, is thus everywhere. One company's onshore is another's offshore. And this powerful world is, in a way, located nowhere as such. Richard Murphy from the Tax Justice Network writes how:

> illicit financial flows ... do not flow through locations as such, but do instead flow through the secrecy space that secrecy jurisdictions create.... They float over and around the locations which are used to facilitate their existence as if in an unregulated ether.
>
> (Murphy 2009: 7)

As a consequence, much of this overlapping 'secrecy world' cannot be regulated by single national states and, indeed, may not be regulatable at all. A world of multiple secrets has developed on such a scale and through an 'unregulated ether' that is almost impossible to tame. The social sciences neglect these offshore worlds at their peril.

Much has been moved offshore – hidden from view, legally protected and not subject to potential democratic oversight, control, and regulation. This offshore world is detrimental to democracy. Many of the quintessential 'offshore societies' are highly undemocratic, often established by a host power to benefit its elites (Britain and the Cayman Islands, China and Macao and so on).

But more broadly such offshoring involves the most sustained of attacks upon governance orchestrated through national states and especially efforts to regulate and legislate on the basis of democratic control. The absolute requirement for good governance is transparency, and that is what secrecy jurisdictions preclude as Keynes warned about in the 1930s (Keynes 1936). These offshored flows of money, finance, manufacturing, services, security, waste, and emissions are bad for such transparent governance. Such transparency necessitates debate and dialogue being able to determine and implement policies through citizens being aware of, and having control over, a clear set of onshored resources. This new order is one of multiple concealments, of many secrets, and some lies. Offshoring erodes 'democracy' and notions of fairness within and between societies through generating a kind of regime-shopping. Thus there is here a major terrain of struggle between offshore worlds and various kinds of democratic and global commons.

Democracy requires activities to be brought back 'home' and the interests of a society's citizens to be regarded as primary. Much needs to be reshored so as to re-establish potential democratic control by the members of a given society over the activities and resources that are specific to them. This is of course a formidably difficult requirement in a global order. One interesting policy is that proposed by the Tax Justice Network (and by the EU) that the taxation of corporations should be based upon treating each corporation as a single entity around the world (Picciotto 2012). Companies would have to submit one set of consolidated accounts and apportion activities to countries according to their actual 'economic' presence within each country. This presence would be formula-based, relating to the number of staff employed, the geographical location of the company's fixed assets, and the value of its sales. Corporation-type tax would then be levied in each country according to this 'country-by-country' reporting for purposes of fair taxation.

Offshoring also precludes the slowing down of the rate of growth of CO_2 emissions which presupposes shared and open global agreements between responsible states, corporations, and publics. Offshoring makes the idea of decarbonising societies hard to realize (see Sayer 2015: Part 5). Powering down to a low carbon future requires a mutual indebtedness of people and especially of current generations towards future generations, including those not yet born. The need for this public or social indebtedness is expressed in many documents, such as the *UNESCO Declaration on the Responsibilities of the Present Generations Towards Future Generations* (12 November 1997). Such social indebtedness between people has been overwhelmed in much of the world through financial indebtedness, which ties people, states, and corporations into financialised obligations. This financial indebtedness and the large-scale offshoring of potential taxation revenue make it exceedingly hard for social indebtedness to gain traction. Public expenditure and a strong notion of a common interest are necessary to plan and orchestrate low carbonism.

Mazzucato (2015) brings out more generally the role of public investment in generating private sector innovations. Offshoring and the effective powering down of economies and societies directly contradict each other. Only a cluster of exceptionally powerful alternative low carbon systems that could offset such tendencies. And this is rendered especially difficult to realize because of offshoring so much relevant within societies, of leisure, work, waste, CO_2 emissions, torture, finance, and especially taxation (Urry 2014).

Moreover, it is possible, as various dystopian futures in film and literature remind us, that we may not have seen anything yet in the scale and impact of offshoring. The twenty-first century could be a century of 'extreme offshoring' with many catastrophic consequences for democracy, post-carbonism, and sustaining any notion of a viable commons. The stakes are high indeed as to how these struggles between offshoring and reshoring various commons will play themselves out in future decades. Reversing offshoring is essential for sustaining most forms of the commons within this unfolding century.

Notes

1 See the video, at: www.youtube.com/watch?v=CChAOh1X5CA (accessed 18 January 2016).
2 See Press Trust of India, 2013, at: www.indianexpress.com/news/high-net-worth-individuals-india-second-to-only-hong-kong-in-growth/1131137 (accessed 18 January 2016).

3 'Revealed: Global Super-Rich has at least $21 Trillion Hidden in Secret Tax Havens', at: www.taxjustice.net/cms/upload/pdf/The_Price_of_Offshore_Revisited_Presser_120722. pdf (accessed 18 January 2016).
4 See: http://opencorporates.com/viz/financial/ (accessed 18 January 2016).
5 See the list here, at: www.actionaid.org.uk/news-and-views/ftse100s-tax-haven-habit-shows-need-to-tackle-a-hidden-obstacle-in-the-fight-against (accessed 18 January 2016); see also: www.guardian.co.uk/news/datablog/2013/may/12/ftse-100-use-tax-havens-full-list (accessed 18 January 2016).

References

Anderson, Jon and Kimberley Peters (eds). *Water Worlds: Geographies of the Ocean.* Farnham: Ashgate, 2013.

Appelbaum, Eileen and Rosemary Batt. *Private Equity at Work: When Wall Street Manages Main Street.* New York: Russell Sage Foundation, 2014.

Bauman, Zygmunt. *Liquid Modernity.* Cambridge: Polity, 2000.

Birtchnell, Thomas and Javier Caletrío (eds). *Elite Mobilities.* London: Routledge, 2013.

Brittain-Catlin, William. *Offshore: The Dark Side of the Global Economy.* New York: Picador, 2005.

Budd, Lucy. 'Aeromobile Elites: Private Business Aviation and the Global Economy', in Thomas Birtchnell and Javier Caletrío (eds), *Elite Mobilities.* London: Routledge, 2013, 78–98.

Davis, Gerald F. *Managed by the Markets: How Finance Re-Shaped America.* New York: Oxford University Press, 2009.

Davis, Jerry. 'Re-Imagining the Corporation'. Paper presented to American Sociological Association, Colorado, August 2012, at: http://webuser.bus.umich.edu/gfdavis/Presentations/Davis%20ASA%202012.pdf (accessed 18 January 2016).

Elliott, Anthony and John Urry. *Mobile Lives.* London: Routledge, 2010.

Farrell, Paul B. 'Rich Class Fighting 99%, Winning Big-Time'. *Marketwatch* 1 November 2011, at: www.marketwatch.com/story/rich-class-beating-99-to-a-pulp-2011-11-01 (accessed 18 January 2016).

Featherstone, Mike. 'Super-Rich Lifestyles', in Thomas Birtchnell and Javier Caletrío (eds), *Elite Mobilities.* London: Routledge, 2013, 99–135.

Haldane, Andrew G. and Robert M. May. 'Systemic Risk in Banking Ecosystems'. *Nature* 469, no. 7330 (2011): 351–5.

Hansen, James. *Storms of my Grandchildren: The Truth about the Coming Climate Catastrophe and Our Last Chance to Save Humanity.* London: Bloomsbury, 2011.

Hayek, Friedrich. *The Road to Serfdom.* London: Routledge, 1944.

Keynes, John Maynard. *The General Theory of Employment, Interest and Money.* London: Macmillan, 1936.

Klein, Naomi. *The Shock Doctrine: The Rise of Disaster Capitalism.* London: Allen Lane, 2007.

Langewiesche, William. *The Outlaw Sea: Chaos and Crime on the World's Oceans.* London: Granta, 2004.

Lash, Scott and John Urry. *The End of Organized Capitalism.* Cambridge: Polity, 1987.

Lash, Scott and John Urry. *Economies of Signs and Space.* London: Sage, 1994.

Mazzucato, Mariana. *The Entrepreneurial State: Debunking Public Vs. Private Sector Myths.* New York: Public Affairs, 2015.

Murphy, Richard. *Defining the Secrecy World: Rethinking the Language of "Offshore".* London: Tax Justice Network, 2009.

Ohmae, Kenichi. *The Borderless World: Power and Strategy in the Interlinked Economy*. London: HarperCollins, 1990.

Palan, Ronen, Richard Murphy, and Christian Chavagneux. *Tax Havens: How Globalization Really Works*. Ithaca, NY: Cornell University Press, 2009.

Picciotto, Sol. *Towards Unitary Taxation of Transnational Corporations*. London: Tax Justice Network, 2012.

Piketty, Thomas. *Capital in the Twenty-First Century*. Cambridge, MA: Harvard University Press, 2014.

Press Trust of India. 'High Net Worth Individuals: India Second to Only Hong Kong in Growth'. *Press Trust of India*, 20 June 2013, at: www.indianexpress.com/news/high-net-worth-individuals-india-second-to-only-hong-kong-in-growth/1131137 (accessed 18 January 2016).

Savage, Michael and Karel Williams (eds). *Remembering Elites*. Oxford: Blackwell, 2008.

Sayer, Andrew. *Why We Can't Afford the Rich*. Bristol: Policy Press, 2015.

Sennett, Richard. *The Corrosion of Character: The Personal Consequences of Work in the New Capitalism*. New York: W. W. Norton & Co, 1998.

Shaxson, Nicholas. *Treasure Islands: Tax Havens and the Men Who Stole the World*. London: Bodley Head, 2012.

Stedman Jones, Daniel. *Masters of the Universe: Hayek, Friedman, and the Birth of Neoliberal Politics*. Princeton, NJ: Princeton University Press, 2012.

Stiglitz, Joseph. *The Roaring Nineties: A New History of the World's Most Prosperous Decade*. New York: W.W. Norton, 2004.

Tett, Gillian. *Fool's Gold: How Unrestrained Greed Corrupted a Dream, Shattered Global Markets and Unleashed a Catastrophe*. London: Abacus, 2010.

Urry, John. *Offshoring*. Cambridge: Polity, 2014.

Walby, Sylvia. *Globalization and Inequalities*. London: Sage, 2009.

3 Politics in common in the digital age

Natalie Fenton

Introduction

The introductory chapter to this volume signals that the commons is associated with collaboration and democratic participation, and concerned with a collective future. Notions of collaborative sharing and political participation are rarely evoked in contemporary times without reference to digital communications whose affordances are said to dramatically enhance both the extent and nature of the collaboration and participation available to the citizenry. The possibilities for a new networked future in an internet-enabled age have been argued not only to democratise political life through a new means of communication, but also to bring forth a shift in the ontology of the political – a new meaning of politics facilitated through enhanced commoning in online spaces. The last decade has seen such politics manifest in public displays of dissent. With uprisings in the Arab world and North Africa against vicious dictatorships; mass protests in Spain, Greece, Italy, and Portugal against an austerity politics that prioritised banks and financial agencies over people and publics; the Occupy Wall Street movement in the USA heralding the rights of the 99 per cent that spread to many parts of the globe; the demonstrations in Istanbul against the urban development plans for a public park; and the protests against racist police discrimination in Ferguson, Missouri, to name but a few.

This chapter addresses political commons as both a concept and a practice that offers a means to challenge many of the assumptions and strictures of (neo) liberal democracy. As a concept it links the commons to Dussel's (2008) notion of critical democracy that far exceeds the limitations of its now defunct liberal namesake. It discusses how the internet has opened up a space by which the sharing of values based on collectivism over competition and the public over profit have inspired protest and mobilisation that at the very least brings the potential for change into our fields of vision. As a practice it considers what we might learn from social movements to date and the many ways they reveal how commoning against the forces of commodification and the injuries of corporate capitalism foregrounds a *resocialising of the political* that places social needs at its core. But it also forewarns against unbridled enthusiasm: we cannot understand the prospects for any political commons if we do not appreciate the multiple ways in which our political lives are curtailed, constrained, and susceptible to domination by 'massive administrative apparatuses, complex markets and the historically powerful peoples and parts of the globe' (Brown 2015: 220).

The internet is a significant part of counter-politics in the digital age: it has galvanised local campaigning and facilitated transnational political movements – bringing forth the *potential* for political commons at an unprecedented level as heralded by Hardt and Negri (2000, 2004) a decade and more ago. However, while it is undoubtedly the case that digital media proffer real advantages for a counter-politics that often circulates in online networks through social media, it can only be understood in relation to what it runs counter to – this includes mainstream politics as well as legacy media. In many Western liberal democracies, mainstream politics is at critical disjuncture. In general, with a few notable exceptions, we have seen a decline in support of establishment political parties and in voting publics (Sloam 2014). In many places around the globe social democratic parties have shifted their policies ever rightwards. In general, in Western democracies party politics have shifted to a 'consumer' style of representation. Both in terms of political parties, who in order to win elections must engage in persuasion, and impression management – what Louw (2005) refers to as 'image making, myth making and hype' on behalf of elite political actors; in terms of the media who, hungry for news fodder, routinely access and privilege these elite definitions of reality and are claimed to serve ruling hegemonic interests, legitimise social inequality and thwart participatory democracy; and also in terms of the citizenry, where being political now includes a vast range of social behaviours (not just campaigning, organising, and argument, but lifestyles, consumer ethics, diet, and musical taste) that has individualised politics and pushed citizenship down consumer-driven channels.

As participation in parliamentary democratic elections declines, so our representative democratic systems face a complexity of problems. These are systems that have been long regarded as weakened by the competitive imperatives of economic globalisation and the declining counter-power of organised labour (Crouch 2004) in the deliberate diminishment of the trade union movement through anti-union legislation. Systems that have endorsed in law the criminalisation of protest and embraced the ever-increasing endeavour to de-public everything from our schools and universities, to our railroads, our health services, and our banks (McNally 2009; Seymour 2014). Systems that suffer from the shrinking range of party political action and conceivable, alternative political projects, and policy interventions (Gilbert 2013); and by the increased power of corporations and transnational financial agencies over public priorities (Mair 2013). Media institutions have been implicated in this pattern of erosion as the subject and object of an economic restructuring that favours elites, both through sustained messages that legitimate the upward transfer and concentration of property and wealth, and through the deregulation and privatisation of the media. These are media institutions so structured and stained by the forces of capital that they are now arguably out of reach of democratically organised political will-formation (Calabrese and Fenton 2015).

All forms of media play a key role in the communication of all things political. But news media in particular are viewed as central to the provision and circulation of information and current affairs and are thus linked to knowledge production and political participation. Yet around the world newspaper circulation has declined sharply and revenue has plunged as advertising has migrated to online sites such as eBay and craigslist. Digitisation has also increased the speed of news reporting, while websites created by newspapers to fend off competition require more space to be filled. News organisations have responded by cutting staff, while expecting those who remain to file more stories in less time resulting in the cannibalisation of news from other online sites. The outcome has generally been that journalism is shallower, faster, and increasingly homogeneous (Fenton 2010). If a politics in common is associated with the concept and practice of democratic politics yet representative democracy and the institutions of mediation that support it seem broken almost beyond repair then the consequences are profound. As Gilbert states, in circumstances where,

> we seem unable to take a decision about what changes to our collective behaviour the situation requires, and to enact them at national or international, or often even at regional and local levels. This then, is fundamentally a democratic crisis: a crisis in the capacity for collective decisions to be taken and upheld.
>
> (Gilbert 2014: vii)

In a period where global capitalism provides, for the most part, the dominant political economic context, we find ourselves facing the extensive failures of contemporary liberal democracy to deliver a workable political system to serve the common good, diminishing the possibilities for a feasible progressive politics in common that could hold the potential to change anything for the better for the majority of people. This resonates with what Colin Crouch has called 'post-democracy':

> the idea of post-democracy helps us to describe situations where boredom, frustration and disillusion have settled in after a democratic moment; when powerful minority interests have become far more active than the mass of ordinary people in making the political system work for them; where political elites have learned to manage and manipulate popular demands; where people have to be persuaded to vote by top-down publicity campaigns.
>
> (Crouch 2004: 19–20)

Bauman and Bordoni's (2014) analysis resonates with conception of a 'post-democracy' – whereby the representative mechanisms that underpin the state have been hollowed out as corporate power and influence has crept in. The state has then abdicated what power it had to counterbalance insidious economic forces and has become 'an institutional idiot' (41), the only role left to it is to attempt to manage public opinion as best it can. Bordoni lists the chief characteristics of post-democracy:

a   deregulation, that is the cancellation of the rules governing economic relations and the supremacy of finance and stock markets;
b   a drop in citizens' participation in political life and elections;
c   the return of economic liberalism (neoliberalism), entrusting to the private sector part of the functions of the state and management services – which before were 'public' – with the same criteria of economic performance as a private company;
d   the decline of the welfare state, reserving basic services only for the poorest, that is, as an exceptional circumstance and not as part of a generalized right for all citizens;
e   the prevalence of lobbies which increase their power and direct policies in their desired direction;
f   the show-business of politics, in which advertising techniques are used to produce consensus; the predominance of the figure of the leader that ... relies on the power of the image, market research and a precise communicative project;
g   a reduction in public investments;
h   the preservation of the 'formal' aspects of democracy, which at least maintain the appearance of the guarantee of liberty.

(Bauman and Bordoni 2014: 140)

All of these factors are interlinked but the first sends shock waves through the rest and leaves us with greatly enfeebled nation states that have little power (and sometimes little will) to deal with the problems they face and democracies that are increasingly bereft. Brown (2015: 108) goes as far as to argue that in neoliberalism, *homo economicus* has displaced *homo politicus* through the insistence that there are only rational market actors in every sphere of human existence such that, the 'citizen-subject converts from a political to an economic being and that the state is remade from one founded in juridical sovereignty to one modeled on a firm.' She argues that the neoliberal response to contemporary problems is framed by markets – more markets, better markets; by finance – more financialisation, better securitisation; by technologies – with an insistence on new technologies and better monetisation of those technologies. It is not genuinely collaborative; it does not involve deliberative and contested decision-making and its democratic implementation through legislative procedures and public policy making; it does not properly plan for the future; it offers no one but elites control over the conditions of their existence.

As neoliberalism subjects all aspects of life to economisation, the consequences are felt not just in the limiting of functions of state and citizen, or in the ever expanding ways in which freedom is defined in economic terms instead of through common investment in public life and public goods, but rather, in the ways in which the exercise of freedom in social and political life is radically reduced and weakened. Recently, Piketty's (2014) study shocked readers through laying bare the ways in which neoliberalism has steered a resolute path to wealth concentration, inequality, and impoverishment unrivalled since the late nineteenth century. Others have also demonstrated extensive and increasing inequality as a neoliberal disease of our own making (Dorling 2014; Oxfam 2015) that massively limits life chances (Wilkinson and Pickett 2009). As inequality increases in neoliberal regimes, so the commons that people share shrink.

But as Skeggs (2014) points out, if we only understand the world from the perspective of the reproduction of the logic of capital, what do we miss noticing or misrecognise politically? We know that there have also been attempts to create spaces protected from colonisation by capital calculation – alternative spaces for progressive political voices such as the Occupy movement, the Black Lives Matter campaign in the USA, the Umbrella movement for democracy in Hong Kong, to mention but a few. Much digital content online also provides the means not only to legitimate, but also to challenge, resist, and counter the hegemony of a neoliberal political-economic order. And so, new forms of circulation of information sometimes constitute alternative bids for political meaning and value (Curran *et al.* 2012), and swimming against the riptide of global capitalism, seek to establish politics in common. Here we find sites where our political lives may be organised differently across space and time (Bennett and Segerberg 2013); where the felt experiences of joining political protest and political groupings online and offline, that seek to counter capital may also convert into the sort of value that encourages appreciation of collectiveness and acting in-common.

The internet and social media have also been criticised for bringing forth an era of easy-come easy-go politics where you are only ever one click away from a petition; a technological form that encourages issue drift whereby individuals shift focus from one topic to another or one website to another with little commitment or even thought; where collective political identity has a memory that is short lived and easily deleted; where googlearchy prioritises commercial sites and concentrates power and authority at 'particular privileged nodes (corporate headquarters, hedge funds, media outlets and so on)' (Gilbert 2014: 22) giving capital ever more structural advantage. Paradoxically, Tufekci (2014) has argued that the lower costs associated with communication and mobilisation via digital connectivity have both empowered social movements around the world and disempowered them by pushing them into the spotlight without the requisite organisational infrastructure for being able to deal with what comes next. So it is undeniable that social media have enabled movements to spread and grow but this has also contributed to a lack of organisational depth – an oppositional politics that is speeded up but spread thin. An overreliance on online organisation is also easy target for repressive regimes keen to censor and surveil the online world for insubordination then:

> seek to divide, polarize and counter its influence by both joining it, with their own supporters or employees, or by beating it, via demonization and/or

bans, which do not completely block motivated citizens but help keep government supporters from using and trusting it.

(Tufekci 2014: 8)

What impact might this have on politics in common when considered alongside:

> the individualizing logic of contemporary consumer culture and much of public policy [that] is indicative of a certain abstract continuity [...] with the forms of individualization typical of industrial modernity and is the first obstacle which any attempt to realize collective power of any kind must overcome?

(Gilbert 2014: 22)

Let's get horizontal

Against the above critical contextual backdrop, it is at least clear that the mediation of politics in common raises crucial concerns that require an interrogation of 'the ways in which the mechanics of neo-liberalism work to inhibit the emergence of any political collectivity whatsoever' (Gilbert 2014: 28). When we are discussing a political commons the many ways in which the 'mechanics of neo-liberalism work' are central to an understanding of the constraints that the political commons must counter. Sennett claims that:

> [t]he new capitalism permits power to detach itself from authority, the elite living in global detachment from responsibilities to others on the ground, especially during times of economic crisis. Under these conditions, as ordinary people are driven back on themselves, it's no wonder they crave solidarity of some sort [...] today the crossed effect of desires for reassuring solidarity amid economic insecurity is to render life brutally simple: us-against-them coupled with you-are-on-your-own.

(Sennett 2012: 279–80)

Sennett argues that the rituals, pleasures, and politics of cooperation provide some of the answers. The ways in which a networked politics has embodied a philosophy and practice of horizontalism speaks to these very rituals, pleasures, and politics. In seeking how to live together better, more cooperatively, to commune against the forces of commodification then a political practice must be agreed upon. The role of horizontalism has been seen as a practical demonstration of the radically democratic values of social movements such as the Occupy movement or the Indignados in Spain. Horizontalism has in turn been seen to be conducive to the networked architecture of the internet. And so, forms of radical politics that circulate via a network of networks, embrace a politics of non-representation, where no one person speaks for another and differences are openly welcomed. Such politics are based on more fluid and informal networks of action than the class and party politics of old. The nature of these struggles resides in the political embodiment of the diversity of social relations they embrace – an explicit contention to resist the perceived dogma of political narratives within traditional leftist politics believed by some activists to be the harbinger of outmoded understandings and values.

In this vein, the forms of radical politics organising online often profess to be leaderless, non-hierarchical, with open protocols, open communication, and self-generating information and identities. Mobilisations facilitated online are said to reflect this fluidity and informality: frequently they display a rainbow alliance of NGOs, new social movements, trade unions, church groups, and a range of political activists from different backgrounds. The differences within and between the various approaches to the politics under contention and deciding upon a unified collective response to a particular cause or concern often raise political dilemmas for activists. They are, however, intrinsic to understanding the vibrancy of a form of politics that prefers to operate with a variety of positions and perspectives as opposed to a traditional class politics of old that may rely on established political doctrines. These networks are often staunchly anti-bureaucratic and anti-centralist, suspicious of large organised, formal, and institutional politics and want to resist repeating what they see as the mistakes of those models in the politics that they practise.

Occupy

There are of course, real differences between as well as within particular movements. Occupy Wall Street was born into a digital age and embodied digital practice. It came into being in 2011 in response to the economic crisis, the bankers' bailout and claims of corrupt political and economic systems that benefit a few people to the detriment of the many. On 17 September 2011 in Liberty Square in Manhattan's Financial District people assembled and then stayed, putting up tents and promising to occupy the space until a process was created to address the problems they identified and solutions were generated appropriate to everyone. By January 2012 it had spread across the USA. Pulling together data from Facebook, news coverage, lists produced by Chase-Dunn and Curran-Strange and the main websites in use at the time, Castells (2015) maps the location of Occupy-related activity in over 1,000 American cities and towns in all fifty states and in Puerto Rico. Occupy also claimed to have generated activities in 1,500 cities globally. The activities were as varied as they were widespread. They involved demonstrations, regular meetings, teach-ins, and occupy encampments with media centres, medical tents, kitchens, childcare, libraries, and even community banks. Calls to demonstrate and join the camps went out on Twitter, Facebook, and on blogs. YouTube and Tumblr were used to share videos and pictures not only of the protest activities but also to tell the stories of those whose lives had been ravaged by debt.

What was so striking about the Occupy movement is that it was a peaceful, collective attempt to deal with failed democracies through coming together on protest camps in public places to publicise disaffection and debate the way forward. This was a desperate search for new political forms of democracy that could be taken out into communities and into society more generally. Developing new forms of democracy is inevitably a painful and difficult task but one that was at the heart of Occupy's practice and wholly embraced through the daily assemblies. It was a practice based on a deep-seated distrust of the political process entrenched in the liberal democratic model that was felt to have turned away from them and towards powerful elites in society. The occupations were organised using a non-binding, consensus-based, collective, decision-making process known as a 'People's Assembly' or 'General Assembly'. People gathered to deliberate and take decisions based upon a collective agreement or consensus. There was no leader or governing body of the Assembly as it operated with the aim of equality of voice. Individual political autonomy was central to the practice of the assembly where the aim was that many singularities of viewpoints could come together in a consensus that was at once expressive of the differences it embodied.

The Occupy London website described the commitment to the consensus process as:

> a living demonstration that each one of us is important. It's a counter to systems that tell us some people count while others don't. In consensus, everyone matters. But for consensus to work, we must also be flexible, willing to let go. [...] Unity is not unanimity—within consensus there is room for disagreement, for objections, reservations, for people to stand aside and not participate.
>
> (http://occupyLSX.org/?page_id=1999)

Because the movement was premised on diversity and inclusivity and its politics rejected the political establishment, it was impossible for it to campaign for specific outcomes. Any one outcome might be unacceptable to another and likely be appropriated into the political mainstream. This form of horizontalism brings with it a commitment (at least in principle) to direct democracy and absolute participation in every decision. The problem is that the principle is one thing and the practice quite another. Direct democracy may be feasible with twelve people or even 120 but with thousands or millions absolute shared decision-making is simply not possible. The Occupy movement was heavily criticised by the traditional Left for refusing to make demands that could lead to some short-term achievable goals that would energise and fuel hope in the movement. Nonetheless, what we saw in Occupy was a cooperative networked politics, an attempt at a politics in common, in embryonic form.

Some of the social movements that have arisen alongside and after Occupy have grappled with the issue of social change and developed their practice via a more politically strategic approach. These movements have understood the need to embrace diversity and to reject a politics of old organised around unifying ideologies such as socialism and communism. In their bid to create a politics in common they have even rejected descriptions such as Left and Right and sought to embrace a politics based not on political ideology but on social need. Elements of these social movements have then taken the bold step of forming alternative political parties based on the premise that if they want to enact a politics of transformation and meet social need on the ground then they must also accept the need to claim power and ultimately to seek government office. Podemos is a case in point.

From the Indignados to Podemos: resocialising the political

The financial crash in 2008 hit Spain hard. Unlike Greece, the Spanish government's borrowing was under control (more so than Germany). On joining the Euro in 1999 interest rates fell, credit flowed into Spanish banks and then into housing creating a massive property boom and construction bubble financed by cheap loans to builders and homebuyers. House prices rose rapidly from 2004 to 2008 then fell when the bubble burst. The construction industry crumbled; over-indebted homeowners faced financial misery and the banks had mounting bad mortgage debts. So even though the Spanish government had relatively low debts, it had to borrow heavily to deal with the effects of the property collapse. Taxes were raised in 2011, a freeze was imposed on public sector pay, and austerity measures were put in place. Unemployment soared to 25 per cent and inequality rocketed with Oxfam reporting that the twenty richest people in Spain had income equal to the poorest fourteen million Spaniards (Oxfam 2014). Politicians were felt to be at the behest of bankers.

The birth of the Indignados/15M social movement emerged from this crisis. It began with a Facebook group that brought together affinity groups such as XNet and Anonymous under the name of '*¡Democracia Real YA!*' (Real Democracy NOW!). On 15 May 2011 they mobilised many thousands of citizens to demonstrate in the streets across Spain. On 16 May people occupied Catalunya Square in Barcelona and stayed for several months to debate issues that were being ignored in the local elections. This triggered similar occupations in over 100 Spanish cities that then spread to over 800 cities around the world (Castells 2015). The movement became known as 15M or the Indignados – it had no formal leadership and was initially largely ignored by the mainstream media. It used the internet to spread the word and people duly came to the squares to participate. But the internet would have made no difference at all had the moment not been right, had the injustices of a failing democracy not been felt, had poverty not been visible; if unemployment had not been a common experience; had political corruption not been rife. 15M ran campaigns against cuts and spawned many protest movements (known as tides) against evictions and home repossessions, the privatisation of health care, cuts in education, wage cuts, and attacks on working conditions among many others.

15M laid four years of foundations for the development of Podemos that enabled the indignation of protest to turn into explicit policies for political change (the name of their first manifesto was '*Mover ficha: convertir la indignación en cambio político*' or Making a move: turning indignation into political change). Podemos is a party against austerity measures with a commitment to a participatory popular politics (similar to Greece's Syriza). Within a year of its formation in January 2014 Podemos (which translates as 'We can'), had become a populist Left-wing party in Spain, had more than 200,000 members and almost 1,000 circles (horizontally organised local meetings), and frequently topped opinion polls. In May 2014, five months after it was formed it gained five MEPs in the European elections (equivalent to 8 per cent of the Spanish vote) with an astonishing 1.25 million votes. Podemos poses a serious challenge to the two-party duopoly that has dominated the Spanish political scene in post-Franco years – the Spanish Socialist Workers' Party (PSOE) and the party currently (in July 2015) in office the conservative People's Party (PP). Although these parties sound different in name, they had become increasingly similar and colluded over a change to the Spanish constitution to take away Spanish workers' rights to appease the IMF and meet the Troika's austerity measures. Podemos promised something different, calling for:

> A fair distribution of wealth and labour among all, the radical democrat-isation of all instances of public life, the defense of public services and social rights, and the end of the impunity and corruption that have turned the European dream of liberty, equality and fraternity into the nightmare of an unjust, cynical and oligarchic society.
>
> (Quoted in Maura 2014)

A large part of what makes Podemos different is where it has come from. It was not a product of the establishment but rather of a social movement; a social movement that saw itself as an embodiment of the political commons. Eduardo Maura, a professor of philosophy at Complutense University of Madrid and an international representative of Podemos, notes that:

> The social movements changed perceptions, they enabled people to recon-ceptualise supposedly individual problems as common ones that demand collective, political responses. Podemos's ability to stand in the European elections was very much dependent on the social power accumulated by the social movements.
>
> (Quoted in Dolan 2015)

In the same interview Maura talks about the need to keep the Movements sepa-rate from the Party to ensure both that the movement remains autonomous and self-regulating and that the Party is made more accountable and has a broader appeal to a wider range of people who may not identify as activists or even con-sider themselves to be on the Left. This sense of reaching out to as wide a con-stituency as possible, of appealing to the unengaged and non-political is linked to the notion of 'popular unity' (Maura, quoted in Parker *et al.* 2014) put forward by Podemos. Popular unity refers to a recognition of the need to create a new common sense to counter the dominant discourses of neoliberalism. This is an attempt at a populist Left politics and in order to gain popularity they have discarded name tags of Left and Right in favour of a discourse of democracy and 'the people' that speaks to the felt experiences of austerity – disenfranchisement, resentment of the establishment, and material deprivation.

Podemos took notice of the horizontality and diversity of the radical politics of 15M and the Indignados movement. They recognised that a politics of everyday life can be more fruitful than the Left's traditional focus on production and Labour (where production and Labour are now so fragmented and insecure that they are difficult to organise around). As Harvey (in Watson 2015) notes, this is politics organised around the spaces that we live in rather than the spaces that we work in. Podemos focus on people's needs and relates improved public services to enhanced democratisation in an attempt to forge a link between the particular and the universal. In this manner they have moved beyond the focus on horizontality and diversity at all costs to recognise the requirement to cohere around a common political goal that can spread solidarity while still being open to debate and difference.

As a party of the digital age Podemos have also adopted what they call a 'hacker logic'. To create a Podemos circle in a local area or on a particular topic requires nothing more than a Facebook account, an email address, and a meeting – there is no membership fee. The principle is to encourage maximum participation from members who then share in the shaping of the development of the party and hold it to account. They rely heavily on the internet to increase levels of participation and accountability through a Citizen's Assembly, although the principles are not so very different from the establishment of any organisation with aspirations of democratic practice. They used the social networking site Reddit for much of the process but also developed apps for voting and establishing agendas. Draft papers were submitted online relating to three areas of organisation, ethics (Podemos refuses any funding from financial institutions and representatives are subject to strict limitations on privileges and salaries), and politics; these were then debated and redrafted/reduced. There was then a period where resolutions were invited on particular topics (as distinct from strategic or manifesto concerns) that were then voted on. Drafts were then discussed at a face-to-face conference attended by 7,000 people that were voted on the following week online. The selection of party candidates to represent Podemos involved live streaming of debates and elections operating on a one-person one-vote system. This process ensured that the most committed activists and those with a vague interest, those who had precious little time to spare, and those without the resources to travel could all take part.

Podemos has however faced stern criticism from activists that it has become a party of the elite and disconnected from its roots replicating the problems of all established parties of old. It is accused of being reformist and of the establishment; of being bourgeois and fake by entering into a state-centred political space that will ultimately reproduce the politics of old. 15M is part of Xnet, a group of activists working for democracy and against corruption that have succeeded in taking over 100 politicians and bankers to court on corruption charges. Largely crowd-funded and aided by citizen collaboration and leaks, it has exposed fraud and financial scams. It is highly critical of Podemos seeing it as very distinct from 15M and part of a 'very old left perspective that is Gramsci centred' and still operates within a traditional political model that is closed to genuine citizen participation (interview with key activist). Their use of the attributes of the internet they say, serves merely to conceal a centralised, male dominated, and egotistical politics. And it is true, that those with substantial cultural capital (notably male economists and academics) have risen to senior positions within the Party. Meanwhile the mainstream media supported by big business paints Podemos as an economic liability and its political representatives as crazed revolutionaries.

What do we learn from this example? The first point is that while the internet can mobilise and involve a lot of people quickly it does not craft a politics or a commons. Although Podemos makes good use of digital media, digital media did not create Podemos. The politics of Podemos arose from a two-party system that had failed the electorate and a programme of austerity that had caused considerable hardship. What Podemos also reveal is that politics is about doing and when we do politics together we create solidarity – this does not mean that we all think of politics in the same way. A contemporary solidarity is forged on the recognition of difference and the constant assertion of that difference through contention but it is solidarity nonetheless. Resocialising the political enables this solidarity to be seen and felt. In a context where corruption has sullied trust; where mainstream media are seen as being de-democratising because they are part of the elite and part of the problem, to be credible and therefore trusted means you have to be not only visible and transparent in your practices but also accountable and inclusive. You have to be able to show that there is another way of doing politics.

Rediscovering the political commons?

Podemos has recognised the importance of the social in the political. The need to come together in a spirit of mutuality and solidarity while respecting differences to deal with the basic problems people are facing on a daily basis – housing, electricity supply, access to welfare, and lack of food. In the face of ever declining respect for politicians, the growing irrelevance of elections as parties increasingly vie for policies that are all too similar and fail to fulfil electoral promises, politics needs to reconnect with the fundamental requirement of the fair distribution of resources. Going into social spaces and special settings, and ensuring basic social needs were met very simply put politics back in touch with the populace. Hence being political stopped being about voting once every few years or signing a few online petitions, it became about doing and being. Digital media may well be part of this doing and being, but they are far from the whole story. Enacting the political commons in this manner should not be confused with the voluntary action of the charitable sector that has long since worked to pick up the pieces where the welfare state has failed or been withdrawn. This is a social provision with a directly political purpose and importantly it reconnects the political with social class and the social realm. This is a politics of democracy that sits closer to what Jacques Rancière (1999) has described as a permanently expansive movement rather than democracy as an institution or regime. The political commons must surely endeavour to be constantly reaching into dark corners that have long since shut out democratic practice.

Civil society is understood generally as the non-institutionalised networks of associative activity outside the state, only here the link back to the state via the will to govern is explicit. Civil society is welcomed by many radical democrats as the site of a new 'post-liberal' democratic politics where new social movements and other self-organised groups spearhead a revival of active citizens in a participatory politics. Instead of focusing politics narrowly on the state or within the institutions of economic production alone – sites that were often presupposed to symbolise the ultimate *unity* of social and political identity – civil society purportedly embraces the ongoing self-construction of democracy and the diversity of identities and struggles within it. Civil society invites an expansion of the political and the displacement of instrumental reason by notions of active participation and deliberation that are indeterminate and undecidable. So we have on the one hand a defunct political democracy and on the other an expansive and diverse civil society. Podemos attempt to bridge the divide between political society and civil society not just by means of a participatory membership but also by virtue of the actual politics espoused to develop a political commons.

It is worth noting that although the political commons may reside in civil society, it is rarely civil in practice. Organisations, groups, and individuals operating in civil society often mobilise intensely held fantasies of social order invested with a profound energy that project contrasting visions of civil life. To preconceive these varied demands as contained, or containable, within an already unified civil space is rather to miss the way they are constituted through contention that opens them to new, potentially subversive, sometimes even violent efforts to redefine the boundaries of social space. For Mouffe (2000, 2005), contemporary democratic theory has a persistent tendency to invoke models of democratic community in which differences are harmonised, and conflict reduced to uncontentious matters of interpretation within a wider context of consensus. Such a view – expressed in 'deliberative' theories of democracy (see Dryzek 2000) – evacuates the political dimension of its conflictuality. That is, it tries to remove the ever-present possibility of antagonism and conflict from political debate. Instead, pluralism is conceived of as a situation in which differences coexist without antagonism.

For both Mouffe and Rancière, it is conflict rather than consensus that defines the political character of a democratic ethos. And it is conflict rather than consensus that characterises both the practice and the experience of radical oppositional politics online. Without conflict, division, agonism, and so on – markers of difference and otherness – democracy loses its function as a practice of regulating difference and collapses into an oppressive homogeneity. By smoothing over differential identities, by appealing to consensus, mainstream liberal democratic theory often narrows down citizenship to those who already agree its parameters – what Badiou (2008) calls parliamentarianism – that is in part what has turned people away from the mainstream political arena. In underscoring the presence of disagreement and conflict, it is possible to promote a democratic ethos that constantly looks to the margins of the public realm to recognise the impossibility of spatial closure or immunisation, of the democratic community from difference. Democracy, in other words, is a condition generated not from the protective enclosure of agreement but, rather, from the very possibilities of conflict brought by our common exposure. A political commons then emerges from contention.

Human beings are also social animals – this is as true for our political lives as it is for any other part of our existences. The social is collaborative as well as competitive; just as the political is collective as well as contentious. In our social domains we take note of people whom we trust. We know from consecutive surveys that trust in politicians is at all time low (Edelman 2015; or see the General Social Survey 1972–2014) and those we trust the most are those in our own networks. Marquand notes that:

> [P]eople are not only happier, but also work better, in more constructive ways, if they feel that their views are fully taken into account. It follows that co-operative, collaborative ways of organizing work are likely to be more constructive, more productive and indeed more innovative than authoritarian top-down systems. It also follows that people, as citizens, prefer to have their views properly discussed. Where there is insufficient consultation, protest is the natural response.
>
> (Marquand 2012)

Podemos may well be mistaken if they think they can win the hearts and minds of the Spanish population solely by reconnecting to the social. If they cannot deliver on their ambitious political pledges, then it is unlikely that the public will continue their support of them. This is not a controllable sphere. But by appreciating the social as the building blocks of the political, Podemos has been able to breath new life into the prospects of a radical political commons to emerge.

Conclusion

This chapter has attempted to address the relationship between the emergence of political commons and digital media. It has argued that political commons are not borne of digital media, although digital media may well facilitate their growth. One of the ways this occurs is through the forms of affective engagement online that offer hopeful surges into the future and connect groups and individuals together in a variety of forms of counter-political protest, collapsing distance and compressing time. Protests contain a mixture of anger, frustration, and sadness amid a cornucopia of feelings of injustice. They may contain moments of joy and elation as well as fear and intimidation. They may change nothing. But they may signal other not-yet realized possibilities that grow in significance as they are shared. Just as the Zapatista movement was said to inspire the anti-globalisation movement that was said to inspire the Indignados in Spain that was said to inspire the Occupy movement that was said to inspire pro-democracy movements across the Middle East. However, when dealing with affect and the connective possibilities of a networked culture we must similarly take account of the very many ways an affective register is called upon to legitimise and sustain contemporary neoliberal politics of austerity through legacy media. There are deep-set structural constraints within neoliberalism that hamper the development of political commons, not least a corporate media that are implicated in the preservation of global capital.

While representative liberal democracy creaks and flounders with diminishing and increasingly disenfranchised voters, it also exposes what a political commons is not. A political commons is not confined to a minority of those who hold political office or circulate around the corridors of power but rather is at pains to be inclusive and open; it is not concerned with sectarian approaches or foundational logics but rather is an open space that is dialogical and dynamic; it will always therefore be agonistic and never be a finished project. Its primary purpose is not to gain and hold onto power but rather to meet social need – it will therefore eschew the conceitedness of unitary party competitiveness to work with all those who share a passion for a more equitable and just society, who desire to live together better. It will relinquish a combative party politics in favour of pragmatic resolutions to social ills. A political commons requires then an entirely open politics with multiple points of access and opportunity for constant participation. To enable the majority to be participants in a political commons it must also be anti-poverty in purpose and against inequality and so must also tackle the accumulation and concentration of global wealth (Harvey 2014). This is no easy task but when democracy has mutated and been debased, reconstituting it demands no less than a collective repoliticisation of the citizenry. I have suggested that social movements of the last decade, enabled but not determined by digital media, can teach us much and are indicative of where a political commons may emerge from.

Resocialising the political is at the heart of this endeavour. Politics has become so disconnected from and distrusted by the majority of people that it needs desperately to reconnect with the social. A process of resocialising involves mutual recognition and information, dialogue and shared radical praxis as these movements slowly and progressively advance a hegemony that includes all demands but may prioritise some. Resocialising the political means just that – political parties that emerge from social movements consist of people who are part of the communities that are dealing with the problems they face through establishing community banks and soup kitchens, tackling corruption, resisting evictions – it is politics with a social practice and one that is without orthodoxy because it is premised on a constant hum of conversation and action that brings experimentation and adaptation so very different from the top-down politics of old. Operationally, this creates the need to formulate mechanisms of genuine citizen participation and control of the spaces we inhabit that can interface with institutional politics – you cannot have mutuality and horizontality without entitlements and regulation. The ordinary politics of daily life must combine with the extra-ordinary politics of institutional change and this will require new forms of state relations that prioritise the value of the public over profit, patience over productivity, and collaboration over competitiveness. This is what Dussel (2008) calls a 'critical consensus' that could lead to a 'critical democracy' based on the real participation of the oppressed and of the disenfranchised as equals in a new political order. A critical democracy will, by its very purpose, call constantly into question the existing levels of achieved democratisation but crucially it will also translate this into an institutional reality. Political transformation is as much about institutional creation as it is about taking power.

A political commons then, must be more than resistance, it must develop an alternative politics that can advance freedom, equality, collectivism, and ecological sustainability while avoiding corporate, financial, and market domination. This will not happen spontaneously but will require organisation that must itself always strive for critical democracy in its own practice – for the symmetrical participation of all those involved. We see the beginnings of this with Podemos who have not only learnt about the power of popular mobilisation and harnessed the support for an alternative social order, but they have also learnt about the difficulties of leading a heterogeneous movement in the face of a viciously hostile mainstream media. The experience of the Syriza party in Greece and its attempt to govern by representing the political choices of its citizens has also taught us about the extra-political power of the banks and financial agencies and that financial power is politically and socially unaccountable. It is on these social and political realities that political commons of the future will emerge though they may not be fully visible yet.

References

Badiou, Alain. 'The Communist Hypothesis'. *New Left Review* 49, January–February (2008): 29–42.

Bauman, Zygmunt and Carlo Bordoni. *State of Crisis*. Cambridge: Polity, 2014.

Bennett, Lance and Alexandra Segerberg. *The Logic of Connective Action*. Cambridge: Cambridge University Press, 2013.

Brown, Wendy. *Undoing the Demos: Neoliberalism's Stealth Revolution*. New York: Zone Books, 2015.

Calabrese, Andrew and Natalie Fenton. 'A Symposium on Media, Communication and the Limits of Liberalism'. *European Journal of Communication* 30, no. 4 (2015): 1–4.

Castells, Manuel. *Networks of Outrage and Hope: Social Movements in the Internet Age*. Second Edition. Cambridge: Polity, 2015.

Crouch, Colin. *Post-Democracy*. Cambridge: Polity Press, 2004.

Curran, James, Natalie Fenton, and Des Freedman. *Misunderstanding the Internet*. London: Routledge, 2012.

Dolan, Andrew. 'Podemos: Politics by the People'. *Red Pepper*, February 2015, at: www.redpepper.org.uk/podemos-politics-by-the-people (accessed April 2015).

Dorling, Danny. *Inequality and the 1%*. London: Verso, 2014.

Dryzek, John. *Deliberative Democracy and Beyond: Liberals, Critics, Contestations.* Oxford: Oxford University Press, 2000.

Dussel, Enrique. *Twenty Theses on Politics.* Durham, NC: Duke University Press, 2008.

Edelman. *Annual Trust Barometer Survey 2015*, January 2015, at: www.edelman.co.uk/magazine/posts/edelman-trust-barometer-2015/ (accessed 18 January 2016).

Fenton, Natalie (ed.). *New Media, Old News: Journalism and Democracy in the Digital Age.* London: Sage, 2010.

Gilbert, Jeremy. 'What Kind of Thing is "Neoliberalism?"' *New Formations* 80, no. 80 (2013): 7–22.

Gilbert, Jeremy. *Common Ground: Democracy and Collectivity in an Age Of Individualism.* London: Pluto, 2014.

Hardt, Michael and Antonio Negri. *Empire.* Cambridge MA: Harvard University Press, 2000.

Hardt, Michael and Antonio Negri. *Multitude.* New York: Penguin, 2004.

Harvey, D. 'Foreword', in M. Sitrin and D. Azzellini (eds), *They Can't Represent Us: Reinventing Democracy from Greece to Occupy.* London: Verso, 2014.

Louw, Eric. *The Media and the Political Process.* London: Sage, 2005.

Mair, Peter. *Ruling the Void: The Hollowing of Western Democracy.* London: Verso, 2013.

Marquand, Judith. 'Economics as a Public Art'. *openDemocracy* 15 February 2012, at: www.opendemocracy.net/ourkingdom/judith-marquand/economics-as-public-art (accessed 4 April 2016).

Maura, Eduardo. 'Europe Needs to Change – and Using Grassroots Democracy is How We Do It.' *Guardian*, 13 October 2014, at: www.theguardian.com/profile/eduardo-maura (accessed 18 January 2016).

McNally, David. *Global Slump: The Economics and Politics of Crisis and Resistance.* Oakland, CA: PM Press, 2009.

Mouffe, Chantal. *The Democratic Paradox.* London: Verso, 2000.

Mouffe, Chantal. *The Return of the Political.* London: Verso, 2005.

Oxfam. *Working for the Few: Political Capture and Economic Inequality.* Oxfam Briefing Paper 178, January 2014, at: http://oxf.am/KHp (accessed October 2015).

Oxfam. 'Wealth: Having it all and Wanting More'. Oxfam Issue Briefing. January, 2015, at: www.oxfam.org/sites/www.oxfam.org/files/file_attachments/ib-wealth-having-all-wanting-more-190115-en.pdf (accessed 30 January 2016).

Parker, Lucy and David Mountain, with Nikos Manousakis. 'Más Allá de la Izquierda y la Derecha? (Beyond Left and Right?) An Interview with Eduardo Maura of Podemos'. *Platypus Review* 72, December–January (2014–2015), at: http://platypus1917.org/2014/12/01/mas-alla-de-la-izquierda-y-la-derecha/ (accessed 16 October 2015).

Piketty, Thomas. *Capital in the Twenty-First Century.* Cambridge, MA: Harvard University Press, 2014.

Rancière, Jacques. *Disagreement: Politics and Philosophy.* Translated by Julie Rose. Minneapolis, MN: University of Minnesota Press, 1999.

Sennett, Richard. *Together: The Rituals, Pleasures and Politics of Cooperation.* London: Allen Lane, 2012.

Seymour, Richard. *Against Austerity: Class, Ideology and Socialist Strategy.* London: Pluto, 2014.

Skeggs, Bev. 'Values beyond Value: is Anything beyond the Logic of Capital?' *British Journal of Sociology* 65, no. 1 (2014): 1–20.

Sloam, James. 'New Voice, Less Equal: the Civic and Political Engagement of Young People in the United States and Europe'. *Comparative Political Studies* 47, no. 5 (2014): 663–88.

Tufekci, Zeynep. 'Social Movements and Governments in the Digital Age: Evaluating a Complex Landscape'. *Journal of International Affairs* 68, no. 1 (2014): 1–18.

Watson, Mike. 'David Harvey: On Syriza and Podemos'. Verso Blog, 19 March 2015, at: www.versobooks.com/blogs/1920-david-harvey-on-syriza-and-podemos (accessed 19 March 2015).

Wilkinson, Richard G. and Kate Pickett. *The Spirit Level: Why More Equal Societies Almost Always Do Better*. London: Allen Lane, 2009.

4   Commons feeling in animal welfare and online libertarian activism

Adam Reed

Introduction: from the textual commons to the felt commonalities of activists

Let me start with a borrowed example of 'the commons' as a historical metaphor at work. In his study of the way British readers invented and elaborated upon literary character, David A. Brewer (2005) focuses on the justifications offered by eighteenth-century proponents of 'afterlife' writing or embellishment (i.e. those invested in narrating further adventures of celebrated fictional protagonists). The vindications exist as defences against possible charges of theft or plagiarism; as Brewer points out, this is a period in which authorial copyright remains contested and the notion of proprietary rights over texts still not fully established (Brewer 2005: 22). At this time, 'the single most readily available metaphor' for thinking about literary property and the non-authorial uses of character is the 'traditional village commons' (Brewer 2005: 11). Indeed, Brewer quotes examples of afterlife writers comparing the reader's connection to literary character with the cottagers' common rights (Brewer 2005: 12). This includes the rights to grow crops and graze livestock on common land, but also, through game laws, to hunt, trap, and eat wild animals. Considered as a kind of 'textual commons', the novel is a space owned by the author but in which readers retain use-rights, including the right to work or develop characters, which thereby escape any absolute authorial claim of ownership.

For Brewer, part of the excitement of the analogy, which never achieved legal authority, lay in what he called 'the underlying economic logic of imaginative expansion' (2005: 12). The idea that readers might retain customary use-rights over literary characters added a new dimension to the already odd property status of common land; in particular, it threw up an obstacle to 'the usual way of thinking about property as the right to exclude the public.' Whether interpreted as a form of 'inherently public property' (Rose 1986), or, like the contemporary legal expert Blackstone as 'a kind of possession without ownership', readers, like cottagers, seemed able to exercise common rights. More than this, Brewer suggests, it was the very exercising of those rights that seemed to guarantee their success; as if 'the mere fact of repeated public use could transform private property into a public benefit' (Brewer 2005: 15). Of course, the obvious differences in the analogy were part of its power. In highlighting the traditional right to use common land, afterlife writers were aware that this was a custom in decline, under threat from enclosures. They were also aware of the differences in economies of scale; whereas cottagers must face the realities of common land as exhaustible material resource, readers, it seemed, might use the textual commons and its immaterial content without fear of depletion (Brewer 2005: 13). Indeed, the *shrinking* of the actual commons appears almost in inverse proportion to the imaginative *expansion* worked upon the novel; an effect Brewer wishes to suggest that continues even after the legal battle over copyright is lost.

In the chapter that follows I want this early instance of the invocation and expansion of the language and practice of commoning put to use beyond its original object of description – literature instead of land – to work alongside my own ethnographic examples, which may also seem at first glance far from the concerns of commoners. Drawn from long-standing anthropological research in Britain, my work here concerns two discontinuous projects: a study of right-wing libertarian – or 'right-libertarian' – activism in London, conducted at the moment their politicking went online, and a study, with ongoing fieldwork, of animal welfare campaigners in Edinburgh. If the prompt for the engagement with the historical example offered by Brewer is a straightforward response to this volume's call to 'see the commons as lost in old shapes and recovered in new forms' (Amin and Howell: Chapter 1), and as a contest and dynamic process from its very inception, the specific rationale for introducing my ethnographic research is much more tangential. In neither of the cases I will explore did I come across 'the commons' as significant ethnographic category, for example, nor did I even see it deployed as an explicit metaphor. In each instance, however, the people I worked with had much to say about the consequences of ownership and the language of public goods, about the concerns of commoning. This chapter then is in part an exercise in deliberately reading 'the commons' *back into* popular talk and action, mobilised around what was, is, or ought to be conceived as *shared*. This includes attention to the issues of expulsion around the condition of property ownership but also the threats of expulsion identified around states of 'public property'. Here, questions we might interpret as about *who* owns *what*, or *who* exercises common rights *in what*, are combined with questions about *when* expulsion or commoning occurs, *who* is owned, and *what* is lost in the 'desirable' urge to enact 'the society of holding things in common' (Amin and Howell: Chapter 1). The idea of a textual commons, including the suggestion of forms of personal possession without ownership, and the transformative capacity of public use, remains a salient counterpoint throughout. But in this three-way conversation, I am less interested in the economic logic behind literary imagination and more focused on the imaginative expansions and contractions assigned to individual and collective acts of both 'having' and 'holding'.

As Brewer goes on to emphasise, the textual commons is not just a metaphor that justifies readers' rights to use or exploit literary characters. It is also an analogy that highlights comparisons between different kinds of 'felt commonality'. Indeed,

Brewer wants to argue that the ties that emerge through the shared use-rights of cottagers to common land can be read as the inspiration for the ties that emerge between afterlife writers, and more broadly between fiction readers, who share a form of use-rights in character. So, for him, a certain 'bagginess' in the definition of who can assert common rights is replicated in the new 'virtual' sociability of those who read the same novel. In fact, unmoored from attachment to specific locales, the felt commonality of readers seems to develop at an accelerated pace:

> readers would invent further adventures for a particular character, see that strangers were apparently doing the same, and so conclude that both they and the strangers were far from alone, that there were still others, yet unknown, with whom they all shared a desire for 'more'.
>
> (Brewer 2005: 14)

In this imaginative augmentation, divergences in the use of character didn't seem to detract from the principle of 'shared affection' for character (Brewer 2005: 21), or from the sense of mutuality that flows from it. These observations also inform this chapter's argument. Like Brewer, I am interested in the various forms of 'felt commonality' – literally the feeling of being in common with others – identified by both the animal welfare campaigners and the libertarian activists I worked with. The juxtaposition of these ethnographic examples develops by paying close attention to the way these discussions appear to coincide with narrated events of both expansion or shrinkage of common worlds. I am interested ultimately in how these 'felt commonalities' emerge, the degree to which they are desired or welcomed, and the reasons why. I consider the ethnographic examples in turn, demonstrating how curious and complex are the forms of commoning they espouse and the commons they create, before highlighting the contradictions and paradoxes in their positions and practices – a recognition of their agonistic politics that I refer to here as 'tragic commoning'.

Back on the land: animal welfare, freedom to roam, and the imaginative expansion of felt commonality

For the animal welfare campaigners I work with, the felt commonality that matters is that between humans and non-human animals. Or, more specifically, it is between humans and other sentient creatures. Like the ethical framework deployed by many animal rights and animal liberation campaigners (see Rigby 2011: 6), their concern focuses on the treatment and living conditions allotted to animals that exhibit signs of consciousness (a division of care that places prime value on mammals, yet that includes birds, fish, reptiles, and a range of other vertebrates). But while the *animal rights* position centres its liberation philosophy on the claim that mammals and other sentient creatures have the right not to be owned or treated as the property of humans, in other words, without any paradox, not to be treated as the 'common treasury' of human beings, the *welfarist* position is far less declamatory and far more pragmatic in fashion. In its campaigns, investigative, and parliamentary lobbying work, the Edinburgh charity I knew did on occasions connect abuse or unethical treatment with the particular dimensions of *property* regimes (for instance, considering a call for the human-household pet relationship to be refigured from a legal status of ownership to one of guardianship or supporting the redesignation of livestock as 'sentient beings' instead of just 'chattels' in EU Directives). However, its members never adopted or deployed the absolute statement that property owner-ship and the felt commonality that they desired (between themselves and non-human conscious-bearing critters) were incompatible. If animal rights advocates typically insist that welfare measures can never be adequate as long as they 'defer to the rights of property owners' (see Braverman 2016), the experience of the charity campaigners was that being owned was not necessarily always the source of the animal protection problem, or at least not straightforwardly.

I want to illustrate the point by looking in detail at the charity's response to one particular piece of legislation: the Land Reform (Scotland) Act of 2003. I have chosen this example because it directly speaks to the campaigners strategic concerns around the welfare of wild animals, not so much through the question of who owns them but through the question of who owns and has the right to use the countryside they inhabit. In obvious ways, the example also takes us back to the issue of 'public property' and to 'the commons', this time not through an export of the term to another domain such as literature but through its apparent return to its original object of description: the land.

As Brewer points out, the idea that eighteenth-century afterlife writers had use-rights in the characters they loved drew an analogy with another con-temporary kind of imaginative expansion: the appreciation and satisfactions afforded by the prospect of the British landscape (Brewer 2005: 13). Much like the common rights of readers, this engagement with 'Fields and Meadows' was seen to invest the observer with 'a kind of Property in everything he sees' (the language, Brewer tells us, belongs to Addison). Once again, this is a possession *without* ownership; exercised through the imaginative use and subsequent enjoy-ment of countryside that is accessible but not actually common land. The emphasis here seems a nice entry point into discussing the Land Reform Act, and the charity's response to it, for the notion that enclosed land can also become the 'Property' of the non-owning individual who enters and observes it, and hence takes on some kind of public benefit, chimes with the language of justi-fication for its enactment. Often billed as one instance of flagship law-making in the then new Scottish Parliament, this Act laid out statutory public rights of use or access to certain forms of private property in Scotland. Introduced in the context of the wider 'freedom to roam' movement, led by outdoor pursuit lobby-ists across Britain, the legislation took on added significance in a devolved Scot-land where the historical enclosure and clearance of vast tracts of land had long been central to political debate about inequality, economic redistribution, and national solidarity. The issue is often summarised in a widely-quoted fact of minority ownership – more than half of Scotland is the personal property of fewer than 500 people (see Wightman 2013). Indeed, the Land Reform Act went a good deal further in opening land to public use than the equivalent law in England and Wales. While not a formal expansion of 'the commons', the legislation did end the landed property owner's historical control over entry. What it principally allowed was 'the right to be on land' and 'the right to cross land', designated as privately owned, for personal or 'recreational' pleasure.

One could well argue that there is an imaginative expansion at play here; by reuniting 'the public' with the land, or at least with a recreational use-right in common, the clear implication is that the people of Scotland will rediscover a *felt commonality* previously eclipsed by the exclusions of private property ownership (see Brown 2014; Vergunst 2013). But my own interest in the Land Reform Act is rather narrower, drawn from the interests of the charity I worked with. To these animal welfare campaigners, the Act offered an expansion of a different order, one which promised to make the protection of *wild* animals on *enclosed* land that much easier. As they regularly pointed out, the bulk of the private land opened-up to public access through the Land Reform Act was countryside given over to what they and other animal advocacy groups termed 'cruel sports': hunting, shooting, fishing, coursing, and stalking. It was the needs of sporting estate owners and managers to organise and control the countryside to meet that purpose that dominated both the manner in which landscape was shaped and the treatment of its sentient creatures. Charity workers are fond of highlighting, for instance, the fact that the distinctively low foliage cover of the Scottish moors is maintained by regular burnings, driven by the requirement to force new heather growth for grouse birds to feed on and to ensure wide, clear-view spaces for the sport of shooting parties. Similarly, woodland on estates is partly planted to encourage predator species for grouse and pheasants such as foxes to live and breed there and hence to make trapping and lamping more efficient. Although the Land Reform Act did not stop landowners managing the estates in this way, it did compel them to allow the public onto them and to permit the pursuit of activities on the land not defined by these ends. For the animal protection charity, the new presence of walkers, birdwatchers, cyclists, canoeists, foragers, and campers on privately owned countryside at least offered the possibility of uses not implicated in target-species control or slaughter. Public access to private land, the compatibility of certain kinds of commoning with private property that we see in Brewer's focus on eighteenth-century imaginative expansion, allowed animal welfarists to identify unexpected opportunities to challenge the culture of 'cruel sports'.

Most strategically, the Act greatly improved the charity's capacity to monitor practices of land management and to witness forms of legal trapping and wildlife crime (see Reed 2016). The organisation's 'field investigator', for instance, could now carry out inspections of sporting estates without the threat of trespass. The right of everyone to be on and to cross private land meant that previously *covert* work, which sometimes involved entering landed property without permission, could now be conducted in the relative open. In the past, and still very much today in England and Wales, this kind of monitoring relied on the calculated use of public footpaths and bridleways, public access channels across private land from which estate and sporting practices could be properly observed. The surveillance of fox hunting, for instance, had often resulted in a legalistic dance with hunt organisers and landowners over the degrees to which individual monitors strayed from these paths. Put simply, the access and recreational use rights assigned to the Scottish public significantly altered the investigator's task; the entry of ramblers and canoeists, campers and birdwatchers meant that land-owners and their managers could no longer automatically or straightforwardly challenge his presence.

The fact that the field investigator can now be on private land without expla-nation may be a convenient outcome of the Land Reform Act, but there is also a sense in which the charity hopes that it leads to a more substantive *solidarity* with recreational users. Casting the investigator as one of many members of the Scottish public on landed property prompted the organisation to consider altern-ative bases for placing them in league. The desire was that these other users might be persuaded to add an element of investigation to their recreation, and hence dramatically increase the number of potential eyes on both legal trapping and animal welfare abuses. As well as encouraging and enabling signed-up sup-porters of the charity to make use of the Land Reform Act to carry out voluntary inspection work on sporting estates, the general goal was to make the rambler more aware and to facilitate the collection of their accidentally discovered find-ings. This includes the setting-up, hosting, and promotion of an online portal for reporting incidents of animals caught in snares or other kinds of estate traps. It additionally involves the production of a 'walker's guide', information put together by the charity, which opens with sentences designed to unsettle the comfortable notion of public access. 'The countryside,' it innocently starts, 'can be a place of tranquillity and is where many of us escape to for some peace, as well as a chance to appreciate the natural beauty of our fauna and flora.' However, the guide continues,

> there is a less attractive side.... Whether it is a stroll through woodland, a hike across the hills or a jog along a country track, it is possible that you may come across a more unpleasant scene than the one you were expecting.

If the observations of ramblers can invest them with a kind of possession in everything they see, this must include, the guide suggests, the prospect of animal cruelty that is inextricably tied to that landscape. The Land Reform Act intro-duced common rights to enjoy privately owned countryside, but it also, from an animal welfare perspective, introduced a common opportunity and common obligation to carry out monitoring on behalf of wild creatures. Crucially, this form of felt commonality relied on a notional expansion, both of the numbers of those able to enter and cross private land and of those willing to report wild animal suffering. It equally relied upon a division between those on the land; the charity anticipated that public access might induce common sympathy among recreational users, based on a newly found antipathy to what it regarded as cruel practices of land management and the abusive behaviour of estate owners.

In this example, then, we see several things happening: the 'imaginative expansion' of felt commonalities, between humans and non-human animals, between animal welfare campaigners themselves, and between those activists and the casual ramblers enabled to access the countryside under new freedom to roam legislation. Moreover, we see that while these commons rights of access offer animal campaigners welcome opportunities to take on 'cruel sports', these forms of commoning are more or less compatible with private ownership of land and with property in general. In these ways, this first ethnographic example points to the complexity of contemporary forms of commoning and the commons.

Network civilisation: libertarian activists, the Anglosphere, and commoning in the name of the anti-commons

While the kind of felt commonality advocated by the animal welfare charity in Edinburgh, based around the fostering of an investigative public sympathetic to the suffering of wild animals, may diverge from the kind of commoning envisaged by the authors of the Land Reform Act, as, that is, a national public of recreational or outdoor citizens (see Brown 2014), in both cases there remains a shared acceptance that opening private land to public use is an unequivocally good thing. In order to throw this assumption into some broader context, however, I want now to introduce a quite different set of ethnographic subjects, advocates for a set of principles, and political ideals dramatically at odds with the notion of public goods and common rights. For the right-libertarian activists I worked with in London, *property* is the sole guarantor of individual freedoms, the basis upon which social rules can be built and a non-coercive public realm imagined. Indeed, far from bemoaning the shrinking commons, these activists *celebrate* the enlargement of property regimes and the general conversion of things held in common into things that individuals have or own. That private ownership, which includes the right to *exclude* others, is central to the kind of minimal statism or even stateless society that they propose (see Reed 2015a). If I asked them, these libertarian activists would be quite horrified by the Scottish Land Reform Act; such an encroachment on the property owner's discretion of possession would for them be emblematic of what they spend their lives campaigning against: the continual unfolding of 'unfree', state-controlled society-making.

But what intrigues me most about these libertarian activists is not so much their counter-factual status as subjects who enthusiastically embrace or celebrate what others regard as a *tragedy* (that is, the ever expanding nature of property regimes); it is rather the articulation of these *anti-commons* views at a moment when the terms and conditions of their own activism appear to be undergoing radical and spectacularly expansive change. Indeed, when I first met them, in 2002, many identified themselves as experiencing a revolution in libertarian practice and advocacy, in large part brought about by the emergence of the internet and the rise of what quickly became known as 'pundit blogging'. As a form of weblog focused on self-editorialising and constant political commentary – this activity combined short, pithy posts, usually no more than 300 words in length, with in-text hyperlinks to relevant media stories; activists were immediately attracted by the innovative possibilities for expressing libertarian ideas and reaching new audiences, new converts. Instead of continuing to write, print, and circulate paper tracts and pamphlets, their critiques and polemics could be published online at the click of a button and be precisely targeted to the news reportage of that day or week. Instead of being read by an audience of a few hundred, predominantly fellow libertarians living in and around the capital city, texts might now be read by great numbers of dispersed, far-flung individuals, and receive almost immediate response through the weblog's 'comments' box. Among those who took up pundit blogging, there was a sense of the ground under which libertarian activism played out irrevocably and quite unexpectedly shifting; 'the future has arrived', I was repeatedly told.

One of the reasons these London libertarians took so quickly to weblogging was that they identified the internet as a new space of *freedom* (see Chadwick 2006; Jordan 2001; Miller and Slater 2000). This is not just because of the lack of censorship and the scope for the unregulated movement of political ideas, but also because of the perceived anarchic nature of the technical standards, which promoted continuous innovations of the form. But while in one view the internet may appear as a new manifestation of the commons, a new form of 'politics in common' (see Fenton: Chapter 3), for my subjects, part of the innovative spirit unleashed by digital technology lay in the opportunities provided for *capitalist* expansion and new forms of *private property*. Not surprisingly, they did not recognise the internet as an expression of common rights or public goods; it was instead a space and a series of activities whose potential lay in the fact it was merely waiting to be owned (this included weblogs themselves, which libertarians quickly asserted as personal possessions, a claim most regularly manifest as the right to exclude certain unwelcome or abusive visitors from leaving comments). Blogging activists told me that the internet and more particularly pundit blogging made sense of libertarianism; digital technology appeared here as its apotheosis. It allowed them to locate or relocate the boundaries and terrain of a 'free' life. At the same time, pundit blogging enabled them to develop or cultivate themselves as 'free' subjects; indeed, they saw it as manifesting libertarian ideals of personal interaction. Crucially, this included the re-cultivation of a libertarian 'social', a new way to register and plot the freely-made associations of the individual subject.

Negotiating this change, once described to me as a switch from interacting with 'a clutch of people in one place' to interacting with 'a thin layer of people scattered all over the globe', became a matter of conscious concern. In searching for a language to account for it, activists soon made appeal to another contemporary libertarian form: the so-called 'Anglosphere'. In fact, one of the ways in which the future was seen to have arrived was precisely in this regard. Pundit bloggers saw their activities as an unwitting instantiation of the practice of free connection (the popular neologism 'blogosphere', rebooted by cyberpunk writer and American pundit blogger William Quick, is a deliberate nod to that suggestion). But the 'Anglosphere' presaged, for its advocates, something more powerfully liberatory. Existing since the 1960s and at this time most publicly propagated in libertarian circles by James C. Bennett, the modish concept of the Anglosphere both advocated and proposed the reality of a 'network civilisation without a corresponding political form' or clear geographical boundaries (Bennett 2004: 80). Grounded in the general constituency of the English-speaking world, membership of the Anglosphere, Bennett argues, is fundamentally defined by individual commitment to the *rule of law* and respect for *contracts* and to the 'elevation of freedom to the first rank of political and cultural values' (Bennett 2001: 1). Libertarian activists in London especially liked the fact that Bennett and others insisted that although not defined by nation states, the 'densest nodes' of this purported network civilisation were to be found in both the United States and Britain, which (for them) had the longest traditions of civil society (Bennett 2004: 80). From there, it is adduced, further nodes of Anglospheric association could be identified in Canada, Australia, New Zealand, Ireland, and South Africa, and beyond that among the educated and English-speaking sectors of regions such as the Caribbean, Oceania, Africa, and India. Here then was an apparent basis for the expansion of the only 'public' that activists acknowledged, a social composed of freely-acting individuals in knowing and consenting alignment.

Part of the power of the language of the Anglosphere for libertarians I worked with was that it combined the promise of decentred, 'free' networks of individuals across the world with the reassurance that their own *historic* connections in the city of London remained valid. Now reconceived as a dense node in the Anglospheric network, London played a role as the host of libertarian activity but also existed as a conduit for activity that travelled or dispersed across the global network. Indeed, activists were keen to stress the manner in which the blogosphere (seen as the virtual embodiment of the Anglosphere) incorporates free and individual subjects well beyond its most dynamic nodes. As one London pundit blogger pointed out to me, the regular contributors on his group blog include libertarians posting from London and the United States, but also someone from Zagreb and someone from Bratislava. 'What surprises me is the extent to which they both effortlessly slot into the vibe and context of what we write,' he reflects, 'I mean I understood Bennett's notion of the Anglosphere well, but here it is actually in action.' Other pundit bloggers I met make the same point in reference to the range of comments their weblogs receive or the new pundit blogs they, as readers in the blogosphere, keep discovering. One activist, for instance, told me about the response to a post complaining about new regulations on the London Underground; among the comments left were responses from readers in Dakar and Kiev. The same libertarian has begun reading and conversing by email with a pundit blogger in Kolkata. 'It's sort of a meta-society, a spontaneous network with no State in the middle,' he enthused. Others celebrate the network by actually posting about the geographical location of the comments they receive or the pundit blogs they read, turning the dispersed nature of production and reception on the blogosphere into new libertarian content.

These relationships between 'scattered' pundit bloggers also provoke wider reflections on the nature of what is held 'in common'. Talking about his communications with the Kolkata blogger, for instance, the libertarian in London began to question the quality of his relationship with other Londoners:

> This is a guy (the Kolkata blogger) whose assumptions about what everyday life should be like are no different to mine; they are certainly no more different than my next-door neighbours in well-heeled Chelsea! And that's one of the advantages of the blogosphere, it filters out the things which separate us, you know a lot of the cultural things, and actually helps us connect.

In this account, the blogosphere *is* the Anglosphere, but it is also *more*. As well as uniting individual subjects through the felt commonality of shared 'Anglospheric' heritage (through, for instance, the British colonial legacy in India), pundit blogging seems to provide a technology for impromptu free association that operates simultaneously by *stripping away* culture or tradition (taking away, in other words, the need for things to be held 'in common' at all). The Kolkata and London libertarians may be the product of an English-speaking world historically committed to respect for property and the rule of law, or imagined to be so, but in the end, and through the filter of blogging, they become revealed at the same time as connected by the natural commonality of their thinking as 'free' subjects. The suggestion here is that the blogosphere can present unmoored individuals to each other without the disguise of 'cultural things'. This is a form of the 'social' that is entirely compatible with the rejection of 'society' familiar in right-wing thinking. As a mode of 'commoning' it is of course correspondingly incompatible with 'culture' conceived of as determining the 'individual', and with the claims of any 'public' above and beyond that 'free' individual's property and self-interest.

As celebrants of the shrinking commons, the libertarians I knew are thus a useful corrective to the notion that 'utopian imaginaries' or dreams of radical publics are necessarily drawn from idioms of collective ownership or common rights. One does not have to endorse their programme to observe that these activists experience an imaginative expansion of the kind of 'social' or 'public' for which they campaign through pundit blogging and the like. Indeed, for them, private property ownership and the new opportunities afforded for ownership online is precisely what makes the properly public *social*. While rejecting the assumption that holding things in common – whether public goods or culture – is either good or correctly identified as the starting point for a better society, there remains the idea that the network civilisation instantiated in the blogosphere will nevertheless improve not only individual but also collective well-being, including the unhindered opportunity for connection. In denying what they read as coercive and still-dominant forms of commoning, these activists believe that they can reach the commonalities that lie, as it were, beyond the commons, in the space and the realm of freedom itself.

Free to roam: tragic commoning in the animal welfare movement

If the disappearance of things once held in common can be a source of imaginative expansion for these pundit bloggers, we might then ask where else might the tragic aspect of commoning lie. The question returns me to the example of the Land Reform Act and the perspective of the Edinburgh animal welfare charity I worked with. For while, unlike libertarian activists, they welcomed the opening of private land to public access, in truth members of the organisation only gave the legislation a muted support. Read as a new form of commoning, the Land Reform Act still appeared to them as severely limited in scope and vision; it was not the collective presence or shared return to the land that they campaigned for or really wanted. Indeed, in some ways the Act exemplified what they identified as the anthropocentric dimension of modern times, the privileging of *human* forms of collective rights and the breakdown of that felt commonality between us and other sentient creatures (see Reed 2015b).

It is precisely this kind of *failure* which we might identify as typical of traditional commoning, which the animal welfare charity wishes to directly address. In its campaign literature against the practice of snaring, for instance, an appeal letter for donations from supporters begins with the invitation to: 'Imagine a countryside where wild animals such as rabbits, foxes, badgers and otters are free to roam, live and play in safety.' The call is grounded in contemporary welfare notions of non-human creaturely consciousness, embodied in the much quoted phrase that 'a sentient animal is one for whom feelings matter' (Webster 2005: 11). The emphasis here is not just on the negative capacity to feel pain but also on the assertion of the positive capacity of individual creatures to be aware and take species pleasure in their environment. It appears that part of that pleasure is now the freedom to range. In this and other campaign literature, the notion that the privately owned countryside has a sinister side is coupled with the challenge to imagine a different Scottish countryside where human and wild animals both enjoy being on and crossing the land. Conceived as an advocate for freedom, the charity seems to want a Land Reform Act that assigns 'outdoor citizenship' for all, a kind of recreational access in common.

This is not a totally fanciful ambition. As the charity members point out, certain species of wild animal in Scotland already receive legal protection, which in essence acts as a form of guarantee of access onto and across sporting estates. Gamekeepers and land managers, for instance, are prohibited from trapping badgers, red squirrels, otters, and pine martens (although, as the charity highlights, the problem of 'by-catch' in estate traps means that non-target species are often accidentally snared). Likewise, it is prohibited to shoot, trap, or poison birds of prey such as hen harriers, red kites, buzzards, and golden eagles. Other wild sentient creatures receive temporary annual protections; brown hares, for instance, may not be taken or killed in the closed season between 1 February and 30 September. Issued in the name of conservation or species endangerment rather than the recreational use-rights or welfare of individual wild animals, these protections nevertheless open a division in the putative freedom to roam. A fox and a pine marten may both cross private land, however, they are not equally vulnerable, at least in law, to entrapment and slaughter. While legislation no longer allows the property owner to exclude humans or certain named species from their land, it does allow them to exclude or control the movement of others.

The anachronistic definition, from the perspective of the charity, of who gets included in 'public' access to private land leads the organisation to take a side-ways outlook on ownership more generally. Support for the Land Reform Act and for arguments in favour of breaking up large sporting estates (quite regularly made in the Scottish popular press) for them comes down to the question of what is the better outcome for wild animal welfare; a hunch exists that division into farmed smallholdings will reduce the degree of animal control and hence the levels of absolute suffering. But there is no ideological commitment to critique land ownership itself. Indeed, as the charity members like to point out, when it comes to the treatment of wild animals there is little necessary difference between their legal position on enclosed, private land and on common land.

By way of elucidation, let us briefly revisit the eighteenth-century idea of the textual commons. As Brewer (2005: 12) highlights, this analogy works in part through the equation of literary character precisely with wild creatures. Characters should be 'free to range', the argument goes, because their immateriality guarantees the same unownable status. But, as Brewer goes on to argue, the freedom of literary character, like the freedom of 'wild beasts', is ultimately upheld for instrumental, human ends. It is precisely the achievement of that non-propertied status that gives readers and cottagers on common land their use-rights. Invoking game laws alongside common rights, the analogy suggests that in their freedom from ownership both characters and wild animals become available for public entrapment. Everyone, or at least those identified with the local commons, is said to gain 'a qualified property through the labour of confining them.' This is not a freedom therefore that assigns rights to the object of property exemption (i.e. the unowned animal); neither immaterial characters nor fleshy beasts have a stake in the commons they roam across. Likewise, on the modern estate and on private land more generally it is not the ownership of wild animals that curtails their roaming (they continue to escape the status of absolute property, to remain *ferae naturae*), but instead the responsibilities to control that may flow from land ownership. In both cases, a vested right of possession only occurs at the moment the creature is trapped or killed.

As the charity points out, the qualified ownership achieved through entrapment or slaughter complicates the question of which legal mechanisms might best serve the interests of the roaming wild animal, to 'live and play in safety'. It is possible that the felt commonality between human and non-human sentient creatures that they wish to promote might best be respected not by removing the right of ownership over them but by introducing new property regimes; the charity notes, for instance, that some states in the USA are experimenting with species conservation laws that actually make certain wild animals the property of that state. At times it appears that the charity is even advocating that the wild animal itself becomes configured as a kind of qualified property owner. Stretching the idea of a multispecies outdoor citizenship to its limit, the perspective allows us to see each snare set for a fox, each funnel or ladder trap set for a crow, each fen trap set for a stoat or weasel as another way of denying not just free movement but also the sentient creature's satisfaction or possession in everything it sees or knows.

Common knowledge: tragic commoning in the Anglosphere

Not surprisingly, the tragedy of pundit blogging is somewhat different. Activists do not suddenly retract or regret the privatisation of public goods or the spread of a properly social libertarian network civilisation. They do not call for, as the animal welfare charity does, an expansion of who gets collected together or included as the recipient of (access) rights and protections. However, activists *do* discover distinct ambivalence in their response to the Anglosphere online. We might read this as a tragedy borne not from a shrinking or from a wrongly articulated commons but from the unexpected and unwelcome return precisely of those things held, managed, and imagined in common. In this regard, the anti-commons politics of the libertarian blogosphere is haunted by the commons it aspires to extirpate.

On the whole, libertarians are philosophical about any negative consequences of a shift from print or pamphlet activism, largely conducted in and around London, to global activism online, mediated through weblogs. They accept, for instance, a certain sense of inevitable marginalisation that comes with entering a dispersed and numerically much enlarged field of interaction. 'Before,' one pundit blogger wryly observed to me, 'I was a relatively big fish in a very tiny pond, the London libertarian scene. Now I am a small fish in a big pond, just another blogger in the blogosphere.' Despite this decentring the activists I knew remained enthusiastic and excited about their place in the new and dramatically magnified network of libertarians. They were considerably less enamoured, though, of anything that felt like a geographic imbalance or national dominance within the network. In this respect, faultlines in the 'Anglosphere', the anglophone world itself being far from a commons of equal access, are obvious.

Since the vast majority of other libertarian pundit bloggers were at that time based in North America (in 2002, the estimated figure was over 75 per cent of pundit blogs), individuals in London found themselves having to quickly adjust to their concerns and routines. In a very basic way, the activist day was now reconfigured to synchronise with transatlantic time zones. The bulk of reader comments and emails on the weblogs, for example, tended to arrive in the late afternoon and evening as first the East Coast and then the West Coast of the United States awoke and came online. This was also the time of day when the liveliest weblog punditry and discussions were published; in order to participate, those in London had to time their posts accordingly. Perhaps more concerning, activists found themselves also suddenly having to address their libertarianism to American questions. To remain relevant individuals had to gain a quick education in libertarian controversies focused on federal or state politics and in issues of national debate such as gun control. Indeed, they began to post their own contributions. While this development could sometimes be welcomed as a feature of the new 'international lifestyle' of libertarian online activism, it could also sometimes be resented. Activists in London complained that debates on the blogosphere were too often 'skewed in an American direction'. This could even be presented as a new form of coercion; 'I feel a bit colonised,' one London blogger admitted, 'yes, I'm tempted to say I get fed up with the domination of Americans.' A suspicion therefore emerged that the supposedly free association of their vaunted network civilisation may in fact mask a mode of external control or forced exchange, the very thing that libertarians struggle and campaign against.

The felt ascendancy of North American interests could sometimes be perceived as a dominance of 'Culture', of the stuff, that is, that the blogosphere was meant to filter away to enable unimpeded communication and connection between individual, free-thinking subjects. As well as the coercion involved in feeling obliged to address one's concerns to American exemplars, this tragic dimension of pundit blogging for libertarians in London is strongly attached to the new and unwanted requirement to provide 'context' for their politics. Individuals reported that for the first time they needed to explain the background to their statements, about London or British political life and current affairs. A weblog post about new health and safety regulations, for instance, now compelled from them an explanation about the Westminster legislative process or the power of local authorities. A post about May Day demonstrations in the capital now drew from them a brief account of the origins of this public holiday. 'A few years ago,' one activist related, 'libertarians in London knew everything in their world. Now you have to explain the whereabouts of things; for example, where Oxford Street is and what goes on there, or even that the Paris referred to in your post is not in Texas.' Once again, the complaint attached to this work of contextualisation is in part that it is not *reciprocated*; libertarian pundit bloggers in the United States do not, generally speaking, feel the need to explain the background to their posts. For London libertarians, this practice risks highlighting not just a division in 'cultural things' but an inequality in the shared knowledge that matters or the stock of things held in common.

The irony of *Culture* resurfacing to confound the ambitions of a network civilisation is not lost on the libertarians I knew. Indeed, they were focused on a further twist to this unwelcome form of commoning. For many, the real tragedy of pundit blogging is that the requirement to make the context of politics explicit seems to also force from them an acknowledgement of what they, in London, hold in common beyond politics. The revelation of common knowledge – where Oxford Street is and what goes on there, why there is a public holiday in May, how the powers of local authorities in the United Kingdom work in conjunction with central government – can seem a lot like the recognition of social or cultural determinants, of things held in common that shape or prefigure the nature of association. In other words, London libertarians in the blogosphere are compelled to confront not just an imbalance in common knowledge but more than that a felt commonality between them that, like all forms of public goods, risks being coercive. It is almost as if they experience a simultaneous shrinkage and expansion of commoning, or perhaps an expansion *through* shrinkage (i.e. their marginalisation in a blogosphere 'skewed in an American direction'). Either way, this is not what they imagined libertarian commoning or the properly social to be; the Anglosphere online seems to generate both a network and a worry about the uncontested nature of things newly identified as shared or collectively owned.

Conclusions

The notion of a 'shrinking commons' implies a historical trajectory of *erosion* (see Amin and Howell: Chapter 1), the accumulative conversion of public goods into private ownership. In this account, common rights or experiments in forms of public property belong in the past. This is something that Brewer himself heralds, for as he argues, eighteenth-century character-reading practices, which present the novel as a textual commons, were quickly displaced by the move to grant authors 'a proprietary monarchic authority over their own creations' (Brewer 2005: 194). The consolidation of copyright, part of a wider growing legal impatience with the whole notion of 'coexistent properties', contributes to the developing assumption that authors have control not just over published text but also over the very capacity to imagine an afterlife for their characters. In such a scenario, the immateriality of character gets re-presented as evidence for single ownership, the absolute tie to the mind of the author (Brewer 2005: 195). Here, the invocation of a textual commons and its practices of imaginative expansion is but a passing moment before that of capitalism; the example of a society that briefly toyed with detaching aspects of the potential of the novel from its original writer, democratising access by assigning readers a qualified property in literary character, has a retrospective enchantment. Indeed, Brewer meant the description to serve as a kind of lament. For him, this is not, self-evidently, how things are now.

An alternative narrative trajectory, embraced by many of the chapters in this volume, documents how citizens of a shrinking commons are fighting back, reclaiming public goods or identifying new ways of holding things in common and resisting their passing over into exclusive property regimes. A study of the right to roam movement in the United Kingdom and of supporters of the Land Reform Act in Scotland might provide one excellent example of this, even of the resurfacing of a notion of coexistent property. Here the landowner continues to own the land but ceases to control access to it, ceding at least some of its uses to a wider Scottish public. So, in some ways, this is a property regime operating outside the 'society of expulsions' (Amin and Howell: Chapter 1), not exactly common land but not entirely singly possessed either. Indeed, the reformed status of privately owned countryside in Scotland elicits a space in which a certain kind of commonality – that of an outdoor or recreational citizenry – can clearly flourish. This is not, however, the story I have told. For the animal welfare charity I worked with, the Land Reform Act and the emergence of the recreational user of private land has been a convenient development in their ongoing task of monitoring wild animal suffering on sporting estates. But it has not been able to address the commonality that counts most to them – that between human and non-human sentient creatures. Among those wild animals for whom feelings, including a feeling for environment, matter, the threat of exclusion continues. Charity members may like to envisage an expansion of the definition of outdoor citizenship to include them, but they do not expect an absolute convergence between this felt commonality and reformed ownership of the land or common rights. In fact, the best option may be to refocus the object of ownership altogether, to explore the potentialities of making wild animals themselves a form of property. It is possible that being owned (by the right owner) may provide wild animals with the protection and access rights to private land that the charity campaigns for and desires.

If the potential of owning sentient creatures can be as important as the question of whose land they are ranging across, then it might be useful to look again at the relationship between felt commonality and private property. As Brewer admits, nineteenth-century readers did not in fact simply suffer the abandonment of their common rights in literary character; they actively participated in what he terms this act of 'enclosure' (Brewer 2005: 194). Indeed, it appears that a readerly imaginative expansion took place through the very insistence on authorial property rights and the future experience of character as the sole labour of its owner-creator. In part, this chapter has been concerned to explore the commonalities that can emerge through the celebration of shrinkage or the conversion of inherently public property into single ownership. And here my second example becomes most pertinent. For libertarian online activists, the blogosphere is no digital commons; in fact, their idea of the network civilisation is grounded in the very principle of individual possession and in an encounter between individuals that is filtered of shared things (such as Culture). The fact that this anti-commoning civilisation does not straightforwardly materialise in the manner they expected for them only demonstrates the ongoing threat from the persistence or emergence of what gets imagined in common.

Moreover, while the animal welfare charity was only really interested in privately owned countryside to the extent that its reform contributed to the ultimate goal of banning traps on sporting estates and hence ending the persecution of wild animals, libertarian activists remained focused on the blogosphere as a space that might still enshrine their dream of 'free' society. Ideally, in fact, some hoped that the problem of common knowledge might be overcome. A few of them even anticipated a future in which the shared context of their own lives in London might be rendered once again implicit. Failing that, activists also began to experiment with the deliberate deployment of context and felt commonality as a strategic resource. They started to notice, for instance, that posts that required background explanation for an American audience might be received without additional context by pundit bloggers elsewhere, for instance, those in Australia,

India, or South Africa. Typical examples included posts about the Westminster system of government, but especially those about sport and in particular cricket – a global game for all that is absent from the USA. In fact, by embracing this shared knowledge and redirecting it across the network, London libertarians found ways to selectively bypass the North American node of the Anglosphere. To some, this felt like a reassertion of freedom and the principle of the network civilisation. Although held in common and hence 'cultural', knowledge of such pastimes had the added advantage that they could equally be re-presented as the outcome of individual enthusiasm. A passion for cricket might be revealed as a singly owned thing, dispersed unevenly across the network but without the conspiracy of Culture.

References

Bennett, James C. *An Anglosphere Primer*. Presented to the Foreign Policy Research Institute, 2001, at: http://explorersfoundation.org/archive/anglosphere_primer.pdf (accessed 18 January 2016).

Bennett, James C. *The Anglosphere Challenge: Why the English-Speaking Nations Will Lead the Way*. Oxford: Rowman & Littlefield, 2004.

Braverman, Irus. 'Introduction', in Irus Braverman (ed.), *Living Legalities: Animals, Biopolitics, Law*. London: Routledge, forthcoming, 2016.

Brewer, David A. *The Afterlife of Character, 1726–1825*. Philadelphia, PA: University of Pennsylvania Press, 2005.

Brown, Katrina M. 'Spaces of Play, Spaces of Responsibility: Creating Dichotomous Geographies of Outdoor Citizenship'. *Geoforum* 55, August (2014): 22–32.

Chadwick, Andrew. *Internet Politics: States, Citizens, and New Communication Technologies*. Oxford: Oxford University Press, 2006.

Jordan, Tim. 'Language and Libertarianism: The Politics of Cyberculture and the Culture of Cyberpolitics'. *Sociological Review* 49, no. 1 (2001): 1–17.

Miller, Daniel and Don Slater. *The Internet: An Ethnographic Approach*. Oxford: Berg, 2000.

Reed, Adam. *Literature and Agency in English Fiction Reading: A Study of the Henry Williamson Society*. Manchester: Manchester University Press, 2011.

Reed, Adam. 2015a. 'City of Purposes: Free Life and Libertarian Activism in London'. *Journal of the Royal Anthropological Institute* 21, no. 1 (2015): 181–98.

Reed, Adam. 2015b. 'Snared: Ethics and Nature in Animal Protection'. *Ethnos*. Ahead-of-print publication, 10 April 2015, doi: 10.1080/00141844.2015.1028563.

Reed, Adam. 'Crow Kill', in Irus Braverman (ed.). *Living Legalities: Animals, Biopolitics, Law*. London: Routledge, forthcoming, 2016.

Rigby, Kate. 'Getting a Taste for the Bogong Moth'. *Australian Humanities Review* 50 (2011): 77–94.

Rose, Carol M. 'The Comedy of the Commons: Commerce, Customs, and Inherently Public Property'. *The University of Chicago Law Review* 53, no. 3 (1986): 711–81.

Vergunst, Joe. 'Scottish Land Reform and the Idea of "Outdoors"'. *Ethnos* 78, no. 1 (2013): 121–46.

Webster, John. *Animal Welfare: Limping Towards Eden*. London: Wiley-Blackwell, 2005.

Wightman, Andy. 2013. *The Poor Had No Lawyers: Who Owns Scotland and How They Got It*. Edinburgh: Birlinn, 2013.

5 The liminal paracommons of future natural resource efficiency gains

Bruce Lankford

Introduction

The study of the commons continues to proliferate metaphors, labels, and concepts better to convey understandings of natural resource and environmental governance. Updating Hardin's 'Tragedy of the Commons' (1968) has produced many examples, including: 'commons and anticommons' where under or over regulation is tested (Brede 2009; Heller 1998); 'inverse commons' (Raymond 2000) where both greater consumption and sharing leads to greater good (as with, say, open source software); 'new commons' (Hess 2008), identified as those without developed rules and institutions; 'invisible commons' (Bruns 2011), covering the specific challenges of groundwater; and the 'semi-commons' where overlapping ownership regimes in water exist (Smith 2008). To this list I have introduced the term 'paracommons', to describe a hitherto neglected commons, that of resources freed up and salvaged by future efficiency gains.

Why is this neologism needed? A changing context of increasing scarcity (whether perceived or real), resource recycling, allocation/reallocation, and so-called green growth establishes new urgencies that drive up an interest in the role of efficiency (Bretschger 2011; Keys *et al.* 2012). The tracking, accounting, and ownership of 'saved', once-inefficiently-used resources will be of paramount importance, as exemplified by Norris's (2011) commentary on a recent water dispute that revolves precisely around these efficiency gains, and to whom these gains accrue: '...the United States Supreme Court's recent decision in Montana v Wyoming brings to the forefront one of the most complicated and contested facets of irrigation efficiency: who owns the rights to the conserved water?' Here, and I will return to this exemplary case, Norris is asking about access and ownership over something – a kind of 'commons' – that has *become* apparent but has *not yet been* claimed: a 'paracommons', in my usage.

Using this idea of a resource freed up in the future, this chapter introduces three key ideas. First, it explores more fully the idea of the 'paracommons' of yet-to-be salvaged natural resource surpluses, losses, wastes, and wastages. The savings of increased resource efficiency can be viewed as a common pool problem, asking the pertinent question: 'in a given socio-technical system (e.g. an irrigated river basin) who gets the material gain of an efficiency gain?'

Second, the chapter examines a defining feature of the paracommons, which is its 'liminality'. Liminality, or 'in-betweenness', exists because any paracommons arises out of the temporal difference between a perceived inefficient *today* and the promise of a more efficient *tomorrow*. Such promissory notes do not of course always hold their value. Liminality accordingly signals the in-betweenness of systems caught between overly optimistic prefigurations of future efficiencies and disappointing or even paradoxical paracommons outcomes.

Third, and closely related to liminality, this chapter explores salvageable resources via an *exteriorising* phenomenon, of 'something-hidden-inside-coming-out'. Part of society's conceptual struggles with waste and wastage is not so much due to their visibility (for instance, as refuse/waste products) as their invisibility or near visibility within the changing consumption of resources driven by societal and environmental trends in turn driven by physical scarcity or new information such as health advice and prices (an example being a reduction in household meat or sugar consumption). What is central here is the manner in which resources potentially surplus to consumption, and previously not even seen as waste or wastage, shift towards being observable and externalised as surpluses to the new lower demand (and thus in the 'new now' as waste).

The chapter briefly concludes on the distinctive technical, ethical, and political problems in governing the paracommons, particularly the distribution of the material gains, drawing attention to: (a) the disparate 'commonist membership' that make up the parties interested in the paracommons; and (b) the powerful advantages held by the proprietor of the socio-technical system making the efficiency gains.

The paracommons

In a resource-scarce world, society is increasingly interested in the efficiency of resource use; how to get more (or the same) from less. The efficient use of natural resources lies behind the idea of a green economy, as for instance embodied in recent initiatives by the European Commission in their Horizon 2020 research programme (European Commission 2014, 2015). The combination of burgeoning demand, fears regarding resource shortages, and the increasing variability of supply as a result of climate change have raised the profile of efficiency within environmental policy (Barrett and Scott 2012). These challenges have expressed themselves in a variety of efficiency and ecological thinking that has arisen in the last twenty years, such as eco-efficiency, industrial ecology, industrial metabolism, and x-factor production (Socolow *et al.* 1996; Reijnders 1998; Schmidneiny and Stigson 2000; Anderberg 1998; CIAT 2012), to which can be added ecological modernisation theory (Warner 2010; York and Rosa 2003), and green growth (OECD 2011).

Concerns regarding efficiency are not new however: Sax (1990) wrote in reference to resource limits on 'spaceship earth';

It is not by accident that we are turning towards the control of waste and water marketing as ways to reallocate existing supplies and meet new demand. There is also increasing interest in reuse of existing water supplies and in technical means to achieve equal output with smaller inputs of water.

(Sax 1990: 258)

Efficient resource use implies that the current level of efficiency is increased to one that is more efficient in the future. In a current or previous state, these losses, wastes, and wastages hold either no value or value to some parties but if salvaged in the future they become assets to new or existing parties. For example, a farmer who consumes less water this year than last year and produces the same yield as the previous harvest is improving both his productivity and 'water-use efficiency'. A necessary condition for an efficiency improvement at a given unit (e.g. field for farm) is that consumption of the natural resource is reduced and this lower consumption allows more of the resource to be 'saved' for other purposes.

For example, when the physical efficiency of an irrigation system goes from 64 per cent to 69 per cent (this is the performance gain), it means that the irrigation system is now consuming five centile units more (the irrigation system 'got' the material gain). However, those five units could have gone to someone or something else. The paracommons is about the competition for the five units 'freed up' by the efficiency improvement.

Thus if we 'save' a resource, what does that mean and who gets the 'saved' resource? In other words who gets the material benefit of an efficiency gain? In my recently published book, this question of competition over resources newly 'freed up' by efficiency gains is considered via the concept of the 'paracommons' (Lankford 2013). Using the example of irrigation, Figures 5.1 to 5.3 introduce how the paracommons arises and how it relates to the commons. All three figures provide a connected sequence and should be read together. Figure 5.1 begins with a common pool in the upper left-hand corner. In this example it is a body of freshwater in a dam, aquifer, or stream. Moving to the next part of Figure 5.1 (upper right), we see four irrigators or farmers competing over this common pool of water, facing competition and rivalry. In the bottom left of Figure 5.1, the farmers' water demands are shown as four segments (A, B, C, and D) of a water allocation pie, leaving some remaining in the common pool. In the bottom right of Figure 5.1, part of these irrigation abstractions have efficiency losses. These losses arise for example via evaporation of water from bare soil instead of through useful crop transpiration. The four farmers each have four different efficiency levels and thus varying sizes of the 'waste/wastage' fraction of their pie segments.

Moving to Figure 5.2, the first part in the upper left imports the final bottom-right 'pie' from Figure 5.1 to continue the story. The net demands (the beneficial crop transpiration part) and the inefficient fractions are teased out into two separate diagrams in the upper right and lower left of Figure 5.2 respectively. Thus we can now begin to see the inefficient fractions more easily as a resource to be 'regained'. Finally, by collating these losses, wastes, and wastages together as

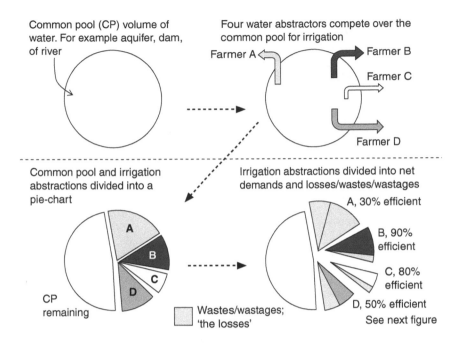

Figure 5.1 From the commons to abstraction and inefficient fractions.

Source: B. Lankford.

one combined segment, we can show in the bottom right of Figure 5.3, how 'the paracommons of losses, wastes, and wastages' begins to arise, showing that it sits alongside the commons (or within depending on how you view it – a point returned to below).

Using the prefix 'para-'

To clarify, while the 'commons' is about competition over existing resources, the 'paracommons' covers competition over salvaged resources from yet-to-be-conserved and more efficiently consumed resources (see Figures 5.3 and 5.4). The prefix 'para-' usefully signals a number of meanings. It indicates first that the paracommons sits *alongside* 'the commons' (typically fish/fisheries in seas or trees in a forest). In this case, 'para-' has a similar meaning to 'parallel': that the paracommons stems from the commons. This new commons arises because parties currently uninterested in the losses locked within the inefficient use of resources are driven by scarcity (and symptoms associated with scarcity) towards saving and freeing-up those assets and then competing over these newly salvaged resources.

'Para-' also has connotations of 'the abstract', as in 'paraphysical'. This is apt since the paracommons is not a physical commons like, for example, fish in a fishery or Brazil nuts in a tropical forest. Rather the paracommons is

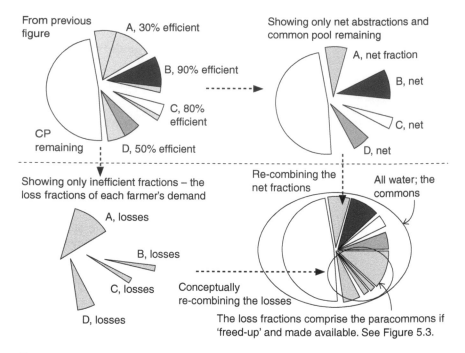

Figure 5.2 Isolating the losses from net consumed fractions.

Source: B. Lankford

pending subject the resolution of where the losses, wastes, and wastages end up. This in turn arises from a human ability to simultaneously observe today's waste and wastage and yet believe, via efficiency improvements, that these losses can be reduced. Once that action has taken place (once the losses have been salvaged) the paracommons condenses or reverts into the commons. This abstract pending characteristic of the paracommons is discussed below under the topic of liminality.

The concept of the paracommons also encompasses 'paradox'. Here, 'para' means 'against' (where 'doxa' refers to 'belief'). The reason that paradox is so central to the theme of efficiency is because without careful planning and forethought, the material gains of the efficiency gain rarely end up where want them to. In other words, the material savings do not return to nature and therefore, paradoxically, do not reduce natural resource consumption. A version of this paradox was described in the nineteenth century by William Jevons (Polimeni *et al.* 2008), but the difference between the two is that price and economics shape the Jevons paradox while resource cascades and pathways during consumption affect the paracommons of natural resources.

Finally, the behaviour of losses over time and space sets up relative and changing viewpoints as in 'parallax'. Put another way, new perspectives on the

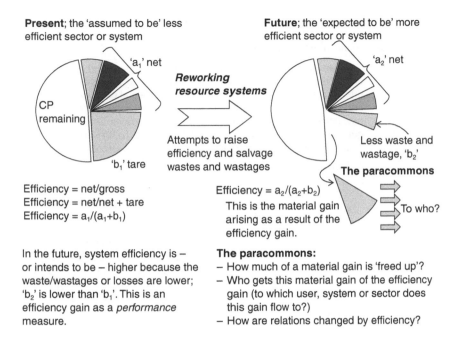

Present; the 'assumed to be' less efficient sector or system

'a₁' net

CP remaining

'b₁' tare

Reworking resource systems

Attempts to raise efficiency and salvage wastes and wastages

Future; the 'expected to be' more efficient sector or system

'a₂' net

Less waste and wastage, 'b₂'

The paracommons

Efficiency = net/gross
Efficiency = net/net + tare
Efficiency = $a_1/(a_1+b_1)$

Efficiency = $a_2/(a_2+b_2)$
This is the material gain arising as a result of the efficiency gain.

To who?

In the future, system efficiency is – or intends to be – higher because the waste/wastages or losses are lower; 'b₂' is lower than 'b₁'. This is an efficiency gain as a *performance* measure.

The paracommons:
– How much of a material gain is 'freed up'?
– Who gets this material gain of the efficiency gain (to which user, system or sector does this gain flow to?)
– How are relations changed by efficiency?

Figure 5.3 Deriving the paracommons by being more efficient in the future.

Source: B. Lankford

efficiency of a system come from those dependent on efficiency changes and salvaging wastes from that system. These changing and alternative perspectives are illustrated in the example of the apple core that follows.

Exemplifying the paracommons via the fate of a household apple core

Figure 5.5 reveals these aspects of the paracommons by looking at the fate of an apple and its 'to-be-discarded' apple core which ordinarily goes to the city garbage, where it might be picked over by people whose lives depend on households throwing away food and goods. Usually for both the apple on a kitchen table and the apple core in the city dump there are two commons: the household members competing over the apple and the city dump 'harvesters' competing over the discarded apple core (e.g. Gutberlet 2008). However, between these two positions are options that might or might not transpire. The household deciding to eat more food (case B in Figure 5.5) or to recycle the apple in garden compost (case C) deprives the waste pickers at the garbage dump of their sustenance. The waste picker fears the outcomes of a household drive to be more efficient and 'green'. Alternatively, a newly installed waste furnace requires waste to generate electricity depriving the householder, the waste picker and the compost of 'their' core.[1]

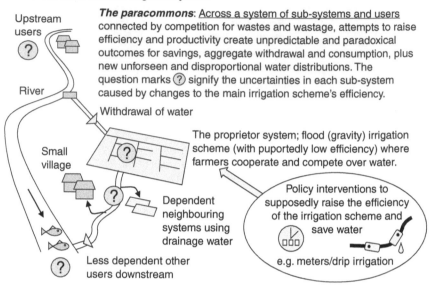

The commons: Within each sub-system, users face rivalry and subtractability over common pool water from; 1) a river; 2) canals within an irrigation scheme; 3) drainage flows recovered from the irrigation system.

Upstream users

The paracommons: Across a system of sub-systems and users connected by competition for wastes and wastage, attempts to raise efficiency and productivity create unpredictable and paradoxical outcomes for savings, aggregate withdrawal and consumption, plus new unforseen and disproportional water distributions. The question marks ⑦ signify the uncertainties in each sub-system caused by changes to the main irrigation scheme's efficiency.

River

Withdrawal of water

Small village

The proprietor system; flood (gravity) irrigation scheme (with puportedly low efficiency) where farmers cooperate and compete over water.

Dependent neighbouring systems using drainage water

Policy interventions to supposedly raise the efficiency of the irrigation scheme and save water

Less dependent other users downstream

e.g. meters/drip irrigation

Figure 5.4 The commons and paracommons of water in a river basin with irrigation.
Source: B. Lankford

Using the apple core example, we can now unpack various aspects of the paracommons.

Rivalrous or agonistic conditions revealed by the apple core exist between four types of groups: householder, garbage picker, furnace operator, and those who speak on behalf of apple tree populations (from seed germination). The four destinations (or dispositions) of the paracommons in the same order are: the proprietor, the usual or immediate neighbour, the wider economy, and the common pool (or environment). Yet the final destination of the apple core depends on switches and changes in interests and technology within the household/proprietor. The city dump, furnace, and compost are all 'downstream' of the household's decision-making.

Furthermore, the four parties only 'sense' or know of each other once changes in the re-routing and amounts of apple core waste begin to happen. Other than that and depending on where the apple core ends up, there are normal 'commons' competing over the waste once it stays or arrives at a given destination (e.g. the apple eaters in the house and the waste pickers in the city dump). Furthermore, these four destinations cannot easily communicate with each other over who gets the core as do the normal or standard commons harvesters inside the house and at the city dump. This disparity means that a 'commonist constituency' is absent or, at best, weak.

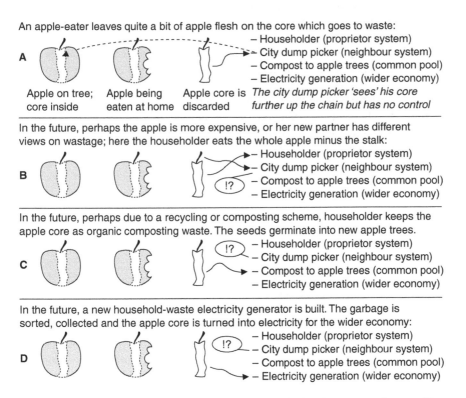

An apple-eater leaves quite a bit of apple flesh on the core which goes to waste:

A

Apple on tree;   Apple being   Apple core is *The city dump picker 'sees' his core*
core inside   eaten at home   discarded   *further up the chain but has no control*

– Householder (proprietor system)
– City dump picker (neighbour system)
– Compost to apple trees (common pool)
– Electricity generation (wider economy)

In the future, perhaps the apple is more expensive, or her new partner has different views on wastage; here the householder eats the whole apple minus the stalk:

B

– Householder (proprietor system)
– City dump picker (neighbour system)
– Compost to apple trees (common pool)
– Electricity generation (wider economy)

In the future, perhaps due to a recycling or composting scheme, householder keeps the apple core as organic composting waste. The seeds germinate into new apple trees.

C

– Householder (proprietor system)
– City dump picker (neighbour system)
– Compost to apple trees (common pool)
– Electricity generation (wider economy)

In the future, a new household-waste electricity generator is built. The garbage is sorted, collected and the apple core is turned into electricity for the wider economy:

D

– Householder (proprietor system)
– City dump picker (neighbour system)
– Compost to apple trees (common pool)
– Electricity generation (wider economy)

Figure 5.5 Who gets the externalising apple core if and when less or more is wasted?
Source: B. Lankford

Critically the common pool nature of the apple core is related to changing perceptions within the household about today's waste alongside 'savings' to be made in the future. Yet the household is in turn reflecting wider societal beliefs regarding acceptable waste and pathways for waste to take. There are both spatial and temporal transitions involved.

Exemplifying the paracommons via case study: Montana vs. Wyoming

The question over who gets the benefit from an efficiency gain was recently writ large by the US Supreme Court decision in 2011 regarding Montana and Wyoming. The Court backed Wyoming's defence that their prior appropriation water law enabled them to use the water freed up by introducing more consumptive irrigation sprinkler systems. The previously 'inefficient' flood technology spilled drainage water that downstream neighbouring Montana had become accustomed to. In the case of Montana vs. Wyoming, the simple expectation was that more efficient irrigation systems would provide more water to downstream

Montana. Paradoxically, it resulted in less water flowing to Montana because Wyoming used the freed-up gain to expand the area under cultivation, resulting in more water evapo-transpired and lost to the atmosphere.

As with the apple core above, the paracommons framework conceives that there are four types of parties or destinations competing over efficiency gains:

1   The proprietor making the efficiency gain (e.g. an irrigation scheme);
2   Immediately connected neighbours (e.g. farmers or villagers using drainage water from the irrigation scheme);
3   The common pool (the river system); and,
4   The wider economy (other users such as industry).

In the Montana/Wyoming case, these respectively correlate to:

1   Wyoming irrigators;
2   Montana irrigators;
3   The Yellowstone River system; and
4   Other economic sectors in both Montana and Wyoming, or further downstream in the Missouri River system.

It is in the light of these competitive parties and forces that Norris' question given in the introduction can be understood: who owns conserved resources and how do we understand these competitive forces and the regulatory environment that arbitrates this competition? Furthermore, this regulatory environment is further undermined by society's inability to accurately monitor resource withdrawal, consumption, and losses at different locations, scales, and times. Clearly though, the key reason that material gains tend not to flow back to nature is because the proprietor and/or their neighbour rapidly appropriate the 'new' resources often backed by legal frameworks ill-suited to dealing with fast-moving and pressing concerns regarding efficiency and scarcity (Neuman 1998;[2] Shupe 1982).

Exemplifying the paracommons with an example from fisheries[3]

One part of managing common pool fisheries might include the establishment of a marine protected area (MPA) where no or little fishing is allowed. The theory is that these benefit fish populations in both the MPA and in a spill-over zone surrounding the MPA. Spill-over is a 'waste' from the MPA point of view but a gain from the point of view of surrounding areas. The phenomenon of spill-over from an MPA to surrounding areas frames the paracommons very well.

Although the paracommons is a relatively abstract idea, as it deals with perceptions of yet-to-be-salvaged-gains, these can be clarified in this example. The term 'salvaged' captures many ways of changing the efficiency of a system and need not mean the active running around with nets trying to grab things. It can also for example mean a change in MPA boundaries so that the newer, larger MPA keeps the spill-over within its new boundaries. So, in keeping with the four destinations or parties identified in my book, these can each get/or share the gain:

1   The proprietor – the MPA. They could do this in two ways: increase the boundaries of the MPA, or put up some kind of barrier (nonsensical to be sure);
2   The immediate neighbour – this would be fishers/ecology within a very narrow strip close to the MPA;
3   The common pool/natural capital – where fish flow to a zone beyond the immediate spill-over zone and therefore replenish stocks in the wider seas (in other words, if there was no or poor immediate neighbourly capture, then the reefs outside the MPA would benefit); and
4   The wider society – via an economically utilised restored reef within the MPA and outside the MPA, giving benefits to other newly arrived fishers, tourists, or food-stocks more generally.

Similar to the examples above, the four different parties involved in deciding on the rivalrous nature of fish spill from an marine protected area may not ordinarily come together to discuss the dynamics and dimensions of their 'common pool'. What would bring together these four parties to consider their paracommons of resources regained would be a programme to reduce waste spillage from the MPA.[4] Unless conjoined over the shifting patterns of the paracommons, they instead remain materially interested in their separated extant commons of fish in their respective domains.

Liminality and the liminal paracommons

Central to the idea of the paracommons are attempts to raise efficiency between a current (and assumed) inefficient system and a future (and expected) more efficient system. These attempts, also aiming to free up a resource, create systems in transition; they put systems on a threshold of change between two states – 'now/today' and 'future/tomorrow'. I have applied the terms 'liminal' and 'liminality' to the paracommons for the manner in which the terms capture the uncertainty and 'in-betweenness' created by multiple options arising when attempting efficiency/productivity changes. Liminality also applies to the idea that losses sit between a state of being salvageable potentially claimed by many parties and then salvaged by a known and specified party – perhaps paradoxically so depending on prior expectations.

The term liminality arose through the social studies of Van Gennep (1909) who explored rites of passage in various societies. The term has, among other applications, described the transitory period between stages of human experience (Buckingham *et al.* 2006), the change within communities (Lawrence 1997), and also in geographical histories of rapidly changing nation states 'being between positions' (Yanlk 2011). In these literatures it is the potential transition-in-waiting, rather than tangible outcomes and new states, that interests scholars.

Liminality is thus very relevant to those studying the promise of a future more-efficient world where we seek to control and place the benefits of efficiency.

How does liminality arise with the efficiency and the paracommons? There are six ways that this question may be answered.

First, efficiency as a performance indicator creates the implicit expectation that performance is to be raised by attempting to rework and improve the technologies and management of systems – though the means and consequences of that may not be fully understood. Murray-Rust and Snellen (1993: 7) in discussing irrigation performance explain: 'Performance indicators, by providing information on past activities and their results, help in making informed judgements which may guide our decision making about future activities.' It is possible to interpret this view more critically; for not only do performance indicators help guide decision-making, they encourage the pre-emptive judging of systems often without measurement (or without accurate measurement), in turn creating the conditions for a perceived efficiency deficit between the system's harshly hypothesised current performance and the future system's expected step-up in performance. This is arguably the story of sustainable development (Hedrén and Linnér 2009) or, more appositely, of Spain's investments in irrigation efficiency in the 2000s – that trustworthy records of existing and future irrigation performance did not exist; instead it was self-evident that a switch from gravity irrigation to drip irrigation would result in the performance gain (Lopez-Gunn 2012).

Second, as Table 5.1 suggests, scientists working on efficiency are unsure about the boundaries in time over which efficiency improvements take place. The gains are transitory (almost ephemeral) because we fail to pin down the start and completion date through which efforts and outcomes can be more exactly tracked and traced.

Third, one can interrogate the implications of the vinculum (mathematical threshold) at the centre of an efficiency ratio. The vinculum is the line or threshold that sits between a denominator and numerator in a ratio. Both in the real world and in the calculation of efficiency, resources in the denominator 'move' across the threshold (or not) to the numerator depending on whether and how benefits are generated.

Fourth, it is important to emphasise that the paracommons is an abstract rivalrous state (as if paraphysical) sitting between a today with real, extant 'commons' features and a transpired tomorrow with its real, extant commons features. Thus, a common pool resource sits as a pre-liminal whole (beneath the vinculum), then is subjected to management and harvest as it 'separates' into many possible streams (with many avoided, forestalled, productive, recovery, and consumptive possibilities) and then re-assimilates post-liminally into a few environmental outcomes or dispositions; depletion, the common pool, or the product (leaving losses beneath the vinculum and 'goods' above it). In this regard thus, the calculation of efficiency contains both useful information but yet also information-loss about today's usage of resources together with a promise of future efficiency gains and movements.

Table 5.1 Different start and end points for comparing efficiency over time

Starting point – the 'today'	Explanation
Design and build	Assumed or designed parameters for efficiency and behaviour are taken from the design specifications. If design assumed very low efficiency then expected gains might be higher than otherwise foreseen.
Average climate; past 'x' years	Records from the last 5–10 years might comprise either real or assumed parameters. For example, if the previous 5 years were reasonably 'wet', and the future time window falls in a dry period, then efficiency gains cannot easily be computed without correcting for moisture differences.
Last year's events	More clearly etched in the memories of farmers, engineers, and policy-makers, the 'today' starting point significantly draw upon recent events such as a drought.
Now, this year	With sufficient measurement, system managers might track current water use and discern current levels of efficiency and losses.
Assumed now	Probably the most likely scenario is to guess the current status of the system without reference to measurement, design, or recent behaviours.
End point – the 'tomorrow'	**Explanation**
Expected future	Policy intended future. This is the prefigured future, without reference to detailed and accurate measurement.
Actual future – end of this season	With sufficient measurement, tracking of losses and efficiency might be possible contemporaneously.
Actual future – a point 'x' years in the future	Again with measurement, it should be possible to generate a post-project evaluation of impacts of efficiency improvement programmes.
Actual average future; average future 'x' years	Ideally impacts should be assessed on the average of a number of years allowing climate variability to be corrected for.

Taking the earlier example of the household apple, one observes a physical transition of the apple core from one location to another, but critically it is the uncertain futures that determine liminality. One compares the reactions of the city dump harvester between her witnessing simply a longer delay ('the discarded apple will now take four hours instead of two hours to make it to the garbage dump') with a new waste policy ('all householders are to throw away less food or to recycle more food waste at home'). The latter is accompanied by much higher levels of uncertainty and worry. So while there is always one apple core, the insertion of an efficiency drive at the household gives the city dump recycler greater concern regarding what is about or not about to transpire. While the apple core remains physically one thing, there are now two expectations for the core: the housekeeper responding to new instructions about waste and the waste picker concerned that less food will arrive at the dump. Previously the unvalued apple core formed only one expectation (or no part of any expectation) as it simply wended its way from the house to the dump as it had many times previously. This dimension of 'expectations' of multiple possible outcomes renders the paracommons liminal.

To put it another way, liminality arises because of the way that humans observe and sense changes in the pathways, amounts and dispositions that can take place alongside changes to resource efficiency. Thus it is via scarcity and the expectation of competition over to-be-salvaged resources that the abstract paracommons arises. But it is because the paracommons revert or condense into the commons once those changes have taken place that the paracommons is a state of in-betweenness between a current normal commons and a future normal commons once a future has transpired (but often not the one expected).

Fifth, liminality arises because, due to the complexity of managing the efficiency of natural resources at different scales, we are not able to fully control how, when and where efficiency within natural resource usage plays out. Although paradox captures the uncertainties of liminality, there is more beyond the 'paradox' characterisation found in efficiency literatures to consider the distributive elements of efficiency via interrelated systems connected by resource recycling and savings. It is these opportunities for competition in the shadow of forthcoming changes, that suggest in-betweenness, transition, and liminality. Liminality produces and is produced by the multiple potentials and uncertainties that sit between policy intentions to raise efficiency and 'real' aggregate outcomes.

Finally, the term 'paragains' (also introduced in Lankford 2013, and defined as material but uncertain gains) signals another kind of in-betweenness, for resource efficiency science does not simply involve a binary distinction between two outcomes: (a) resources are always freed up by efficiency gains; and (b) no resources are freed up by efficiency gains. These two camps have their respective protagonists – for example, see the recent heated debate in the journal *Water International* (Gleick *et al.* 2011). In the pages of this journal, two separate camps went head to head, with one party, led by the Pacific Institute arguing that spare water for California was to be found in currently inefficient irrigation systems ('the next million acre feet' went the future framing headline). An alternative view (see, for example, Frederiksen *et al.* [2011, 2012]) was that no such water existed 'spare' as it was already being recycled via nature through aquifers and drainage lines. This example tells us that the paracommons is about the hopeful expectation that considerable spare water exists within inefficient use for future allocation provided it can be managed, 'freed-up', tracked, accounted for, and delivered to the desired destinations.

Exteriorising: an inside-resource-coming-out

The third section of this chapter argues that the complex nature of paracommons stems from waste/wastage-type resources which, in the face of natural or artificial scarcity, are undergoing value judgements regarding their degree of wastefulness and how this then has a bearing on the visibility of that 'waste/wastage'. Returning to the example of the apple core, we see this very clearly. First, people who eat the apple core without thinking/blinking do not see the apple core as waste; they see the core as food. This contrasts with people who eat the apple flesh but always leave the apple core as waste/wastage. In this latter case, the waste/wastage of the apple core is immediately visible. It is less visible in the case of the person who eats the whole apple.

But these differences in individual behaviours gives us an insight to the second, more important, insight about at which point or stage the apple core becomes visible. For the householder who does not eat the apple core but discards it, nevertheless in the supermarket is 'only' buying an apple – in the act of buying she or he does not see the apple flesh and apple core as two separate items. Similarly, a person well-used to eating the entire apple except the woody stem does not in the supermarket 'see' or buy the apple core as a separate item. For both types of apple eaters, the apple core in the supermarket (further up the chain than the household eating patterns) is hidden and does not feature in the purchasing decision.

Now we get to the third and most important point: the householder who changes her ordinary practice of throwing away the apple core to eating it (or putting it in the compost, or putting it in the bin going to a biomass incinerator) is adding to the visibility of the apple core. For now the ordinary pathway of the apple core is disrupted. The waste pickers at the dump are consequently bereft of 'their' resource. Thus paradoxically, with their survival and livelihood in the balance, it is the city dump harvesters who are more mindful of the apple core's existence, even to the extent of realising its presence prior to its manifestation as a part of the consumption of the apple flesh. In other words, of all the parties, and precisely because of the risk of changes to the routing of the apple core, it is the dump harvesters who are most aware of the apple core sitting in the supermarket.

Drawing on Strathern's interest (1998, 2000) in fluid boundaries between the internal and external regarding the metabolism of villages within Papua New Guinea, I consider that the paracommons *exteriorises* and then valorises previously hidden wastes and wastages under such changing circumstances. In fact, in Lankford (2013) I argue for the word 'tare' or 'loss' rather than wastes and wastages because the latter two words describe resources that are already recognisable or have become visible. The predominant literature on the topics of common pool studies of waste materials rejected by urban populations (Gutberlet 2008) and recycling in the study of industrial ecology (Bourg and Erkman 2003) simply cannot comment on this aspect of externalising what will become 'losses' because they are already dealing with and then valorising extant waste materials.[5]

There are a number of mechanisms by which resources reveal or make tangible their inner or hidden loss (a topic treated in more detail in Lankford 2013). First, losses can be identified and then forestalled or recovered. For example, the non-beneficial evaporation from standing water in an irrigator's field can be forestalled in the following season by either reducing the amount of water added to the field or recovered by adding drains to drain off the excess water. More significantly, society can also decide to free up a gain by shifting the consumption of a current resource to a surplus 'loss'. Put another way, society can forestall and forego consumption, and it can do this in three ways:

1   By de-materialising a resource. Dematerialisation takes place via the reduction in size, density, and weight of goods which then gives rise to a reduction in net demand (or amounts of material consumed in per unit production). Using bread as an example, smaller loaves of bread exemplify weight dematerialisation and less nutritious or less dense bread represent density dematerialisation.
2   By retrenchment of a resource. Retrenchment results in the decrease in consumption of the number of units or goods. An example is the reduction in the number of bread loaves or bags of sugar purchased and eaten within a household or village or other unit of interest.
3   By substitution of a resource. Here one resource is swapped for another. A number of approaches exist depending on which part of the chain of production is substituted and effects can be very similar to dematerialisation and retrenchment. An example is when bread is substituted for rice crackers, or when wheat flour is substituted for rye or bran, or sugar is swapped for sweeteners.

Taking the effects of all three options together (substitution, retrenchment, and dematerialisation), we can work through an example of the conservation of wheat flour to make loaves of bread. Say a head of a household that previously used 3 kg of wheat per week to bake bread now only uses 2.5 kg. Thus 0.5 kg wheat has been 'freed up' and made exterior to their core need (which has changed from 3.0 to 2.5 kg). The paracommons argument says that the destinations for the saved 0.5 kg comprise: (a) the household head (proprietor) who then bakes, for instance, pies with the left over wheat; (b) her children (the immediate neighbours) who bake, for example, cupcakes using the wheat; (c) the wider economy, for the wheat remains in the supermarket bought by other housekeepers; and finally (d) nature, as the 0.5 kg wheat is not grown in the first place and so nutrient depletion from farmers' fields is reduced.

This example also tells us about how wheat external (the 0.5 kg) to the new need (2.5 kg) becomes visible. While this example might seem clear enough, one has to understand that this 'making visible' sits alongside many other factors and drivers regarding consumption, price, and existing wastes. Moreover, the English language deals poorly with this process of exteriorisation. While we readily have words and terms for the wastes and wastages, such as crumbs on the chopping board, flour spilled during baking or as dried crusts thrown away at the end of the week, we struggle with terms and expressions to capture these 'newly exteriorised excesses regained' (a convoluted but accurate term, which makes the point about the limitations of our language and conceptualisation).[6]

Discussion: society and the commons

Evolving commons thinking provides the prism through which to see the richness of nature–society. Examining enclosures and exclusion has done much for opening up inclusionary and encompassing deliberative and metaphysical spaces (Amin and Howell: Chapter 1; Cutcher-Gershenfeld and Lawson 2015). The paracommons introduced briefly via the ideas discussed above provides another means by which we can comment on this evolving space. Four ideas may be selected, starting with complexity.

Commons complexity

First, the paracommons emphasises an emerging and increasing complexity of the commons – a topic that is recognised within the commons literature (Berge and Van Laerhoven 2011) including this edited volume. In other words, the boundaries, content, and cross-linkages of the commons are increasing. Research on environmental issues increasingly recognises that complexity is a defining character of the sustainability, conservation, and governance of common pool resources (Manson 2001, 2008; Scoones 1999; Underdal 2010) and that, if anything, this complexity appears to be increasing not only through more refined understandings of nature–human interactions (Norgaard 2010) but because scarcity, innovation, and rising population disturb the balance of environmental protection and economic development (Tainter 2011).

Clarifying reasons for rebound; the utility of efficiency

The paracommons forewarns us that greater efficiency can lead to a 'rebound' of greater consumption (Polimeni *et al.* 2008). However, the 'paradox of rebound' envisaged by Jevons occurs for different reasons. The Jevons rebound takes place because of the changing prices, economics of production, and consumption driven by efficiency. On the contrary, a more complete expression of the paracommons, exemplified by irrigation, apples, and bread, deals with the changing and nested material pathways of 'losses' flowing to different pathways destinations when the system undergoes efficiency changes. An efficiency rebound of water higher consumption in river basins (Crase and O'Keefe 2009; Ward and Pulido-Velázquez 2008) takes place because of the poorly controlled materiality of efficiency.

Nature–society boundaries and directionality

The paracommons also throws light on a critical revisiting of the modernisation of nature started by Hays in 1959 and continued by Clark and York (2005) and others: that one of the great aims of environmental governance is to protect and sustain nature while meeting the economic needs for the growing human population of the planet. This implies a building up of the comprehensiveness with which we govern the commons utilising resource efficiency and its implied technologies and institutions. There is much that can be said here on the problematic mechanisms for achieving sustainable environmental governance such as green capitalism. By problematising efficiency and productivity within this modernisation trajectory, the paracommons points to profound and multiple sources of 'systems uncertainty'; that we are no longer certain what the system is, who represents it, and how and in what direction resources flow. In contrast to the (in my view) conventional framing of nature-serving-society as through an ecosystem services prism, it is possible to see that resource and waste/wastage flows create a complicated nested and recursive embedding of social-ecological-technological systems. In these systems, resources and paragains flow, cascade, interact, and switch; are attractive to some and neutral or harmful to others; and are in quantities and qualities that change rapidly over time and space.

In contrast to a linear interpretation that views services flowing from nature to a system of use (which I believe is underlying message of the Millennium Ecosystem Assessment [see MEA 2005, Figures A and B]), in an increasingly scarce and populous world, resources are recovered, forestalled, and competed over in the shadow of efficiency-driven redistribution. A more cyclical industrialised yet embedded model of nature may be the more appropriate. In short, it is my view a linear ecosystem services interpretation, while acknowledging complexity (Ruth *et al.* 2011; Norgaard 2010), does not fully speak to the reciprocity between realms and redistributions of resources, surpluses, losses, wastes, and wastages.

Going further, if industrial ecology covers the web of interactions between industrial units, then an increasing 'industrialisation' (for reinventing of nature, see Banerjee 2003; for ecological modernisation, see Bailey *et al.* 2011) of the natural world arises from society's rising interest in the reuse previously discarded or minimally regarded resource wastes. Livestock excreta used as biogas and fish waste used for animal meal exemplify the shifting patterns of consumption and recycling in response to supply, demand, and ingenuity but nonetheless are extant waste products. On the other hand, the paracommons is interested in society's as yet unrealised savings and considers this ever tightening cycle between consumption and re-consumption to the extent that nature's provision of resources is increasingly a function of how humans cascade resources within society.

Fluid fugitive resources; outpacing regulatory instruments

Resource users might also observe that normal regulatory instruments (legal, market, and customary) for the ownership and regulation of consumption, allocation, recycling, and losses are being outstripped by fast-moving events driven by scarcity, necessity, and technical ingenuity. This is one way of interpreting the escalation of the Montana vs. Wyoming case to the US Supreme Court – that prior appropriation US water law seemed unfit to adjudicate upon the question of who owns regained losses.

If not regularly updated and reformed or dealt with via a more relaxed hybridisation of approaches (Smith 2008), normative procedures for managing resources also sitting within disciplines such as engineering and law will fail to advance equitable allocation and productivity gains while attending to environmental, social, and sustainability criteria. With this regulatory 'lag' in mind, Hooper and Lankford (2016) discuss how resource allocation is perhaps more a function of hidden influences than of standard and normal tools.

Outmoded or fixed instruments for governing resource consumption and allocation may additionally not help if we become overly interested in more rosy future scenarios driven by expectations and assumptions (or 'the political economy of promise': Leach *et al.* 2012) more than the governance space of today or of the transition itself between the today and the tomorrow. This desire to consider 'futures' as an alluring topic of study over the messy today characterised by a multi-spectrum mosaic of actors, scales, technologies, institutions, and interventions (Halsema and Vincent 2012) is a risk highlighted and forewarned by the liminal nature of the paracommons.

Conclusion

Echoing the agonistic nature of the commons, the paracommons is an idea describing the competition over future resources 'freed up' by efficiency gains in the face of or driven by increasing scarcity. The paracommons argues that 'savings' of the inefficient part of resource use become a matter of competition between stakeholders found in four different types of destinations. Furthermore, by using this term I hope to capture the sense that resource efficiency is highly complex and that attempts to create more efficient socio-technical systems result in unexpected paradoxical outcomes.

As well as this liminal character of today's wastage and waste being reduced in the future, a closely related feature is the manner in which a reduction in consumption frees up resources ordinarily not yet seen as waste or wastage. The future continuously externalises the 'tare' and it is this increasingly large 'to be regained' resource that is then subject to competitive interests. The paracommons can thus be seen as a 'politics of externalisation and expectancy' (or 'of promise' – Leach *et al.* 2012) framing the uncertain differences between: (a) the prefigurations of the promise of efficiency and productivity gains; and (b) the extant and often unforeseen material, productive, and distributive physical and social outcomes for users and resources following attempts to make savings and reduce consumption of natural resources.

At a more conceptual level, and how rethinking the commons may tell us about nature–society and questions of resource justice, the concluding message of the paracommons is that resources found by savings and efficiency gains present a highly unusual 'commons', and for three reasons. First, efficiency gains are rivalrous but between parties that ordinarily do not meet in the village, ocean, field, or forest. Simply put, commonist membership is highly disparate. Second, the gains are subject to what a proprietor of a particular system does, being particularly advantageously placed to appropriate those gains – in turn aided by legal, economic, and technological doctrines better suited to a standard commons model. Third, the material gains are and will be 'liminal' and hidden, pending the resolution of many different context-specific factors, such as technology, terminology, social learning, scale, and measurement (to name but a few). In summary, achieving a more equitable distribution of efficiency gains between the four separate parties/destinations will present extremely difficult moral, ethical, and political questions requiring perhaps new and governmental regulatory frameworks especially since a communal solution is unlikely (because there is no community in the ordinary sense). A fundamental rethink of current governance instruments (such as water rights) which sustain the advantages of the proprietor may be needed if society is to redistribute material efficiency gains out of the proprietor's existing 'enclosure' to immediate neighbours, the wider economy, and to nature.

Notes

1 That the establishment of a waste incinerator plant deprives waste dump pickers of their livelihoods, see: www.guardian.co.uk/world/2012/jul/02/future-for-india-waste-pickers (accessed 18 January 2016).
2 Neuman contends that entrenched water abstraction law (and water markets) in western USA: 'has revealed itself to be woefully inadequate at eliminating waste and encouraging efficiency. Beneficial use affirmatively protects inefficient water use customs and practices' (Neuman 1998: 996).
3 With grateful thanks to Eny Buchary (Stockholm Resilience Centre) for her part in drawing my attention to this system and for assisting me in explaining this correctly.
4 In a natural environment such as an oceanic reef, I do not envisage such a programme being economic.
5 Much of the valorisation thinking around waste describes dealing with products that have little value at present but which can be made more valuable (Luque and Clark 2013). This is not the same as dealing with an exteriorising process that brings to the foreground materials not yet witnessed as waste.
6 This lack of ready terminology remained a problem in the writing of the Lankford 2013 book – hence use of the words 'tare' and 'paragains'. In an email to me with reference to my conference presentation in Cambridge (email dated 11 September 2014), Marilyn Strathern wrote:

> I still think the idea of waste not as some kind of surplus or excess but as an irreducible internality (aka 'externality') is thought-provoking. When I went back to the Mekeo material I realised, though, that probably the most interesting lesson to get from it is that we have to stretch our language – as you do with the term paracommons.

References

Anderberg, Stefan. 'Industrial Metabolism and Linkages between Economics, Ethics, and the Environment'. *Ecological Economics* 24, no. 2–3 (1998): 311–20.

Bailey, Ian, Andy Gouldson, and Peter Newell. 'Ecological Modernisation and the Governance of Carbon: A Critical Analysis'. *Antipode* 43, no. 3 (2011): 682–703.

Banerjee, Subhabrata Bobby. 'Who Sustains Whose Development? Sustainable Development and the Reinvention of Nature'. *Organization Studies* 24, no. 1 (2003): 143–80.

Barrett, John and Kate Scott. 'Link between Climate Change Mitigation and Resource Efficiency: A UK Case Study'. *Global Environmental Change* 22, no. 1 (2012): 299–307.

Berge, Erling and Frank Van Laerhoven. 'Governing the Commons for Two Decades: A Complex Story'. *International Journal of the Commons* 5, no. 2 (2011): 160–87.

Bourg, Dominique and Suren Erkman. *Perspectives on Industrial Ecology*. Sheffield: Greenleaf Publishing, 2003.

Brede, Markus and Fabio Boschetti. 'Commons and Anticommons in a Simple Renewable Resource Harvest Mode'. *Ecological Complexity* 6, no. 1 (2009): 56–63.

Bretschger, Lucas. 'Sustainability Economics, Resource Efficiency, and the Green New Deal', in Raimund Bleischwitz, Paul J.J. Welfens, and ZhongXiang Zhang (eds), *International Economics of Resource Efficiency: Eco-Innovation Policies for a Green Economy*. Springer: Heidelberg, 2011.

Bruns, Bryan. 'Crafting Rules for an Invisible Commons: Responding to Yemen's Groundwater Crisis'. Paper presented at the University of North Carolina, Chapel Hill, 17 March 2011.

Buckingham, Susan, Elodie Marandet, Fiona Smith, Emma Wainwright, and Marilyn Diosi. 'The Liminality of Training Spaces: Places of Private/Public Transitions'. *Geoforum* 37, no. 6 (2006): 895–905.

CIAT. *Issues in Tropical Agriculture Eco-Efficiency: From Vision to Reality*. Columbia: International Center for Tropical Agriculture, 2012.

Clark, Brett and Richard York. 'Carbon Metabolism: Global Capitalism, Climate Change, and the Biospheric Rif'. *Theory and Society* 34, no. 4 (2005): 391–428.

Crase, Lin and Suzanne O'Keefe. 'The Paradox of National Water Savings'. *Agenda* 16, no. 1 (2009): 45–60.

Cutcher-Gershenfeld, Joel and Chris Lawson. *Valuing the Commons: A Fundamental Challenge across Complex Systems*. NSF/SBE 2020 White paper on Future Research in the Social, Behavioral and Economic Sciences, 2015, at: www.ideals.illinois.edu/ bitstream/handle/2142/17422/Valuing%20the%20Commons%20white%20paper%20 Sept%2029%202010.pdf?sequence=2 (accessed 18 January 2016).

EC [European Commission]. *Horizon 2020 Societal Challenge 5: Climate Action, Environment, Resource Efficiency and Raw Materials*. Orientation Paper, European Commission, 2014.

EC [European Commission]. *Horizon 2020 – Work Programme 2014–2015: Climate Action, Environment, Resource Efficiency and Raw Materials. Revised*. 2015, at: https://ec.europa.eu/research/participants/data/ref/h2020/wp/2014_2015/main/h2020- wp1415-climate_en.pdf (accessed 18 January 2016).

Frederiksen, Harald Dixen and Richard Glen Allen. 'A Common Basis for Analysis, Evaluation and Comparison of Offstream Water Uses'. *Water International*, 36, no. 3 (2011): 266–82.

Frederiksen, Harald Dixen, Richard G. Allen, Charles M. Burt, and Chris Perry, 'Responses to Gleick *et al.* (2011), which was itself a response to Frederiksen and Allen (2011)'. *Water International* 37, no. 2 (2012): 183–97.

Gleick, Peter H., Juliet Christian-Smith, and Heather Cooley. 'Water-Use Efficiency and Productivity: Rethinking the Basin Approach'. *Water International* 36, no. 7 (2011): 784–98.

Gutberlet, Jutta. *Recovering Resources – Recycling Citizenship: Urban Poverty Reduction in Latin America*. Farnham: Ashgate, 2008.

Halsema, van Gerrado, E. and Linden Vincent. 'Efficiency and Productivity Terms for Water Management: A Matter of Montextual Relativism Versus General Absolutism', *Agricultural Water Management* 108, (2012): 9–15.

Hardin, Garrett. 'The Tragedy of the Commons'. *Science* 162, no. 3859 (1968): 1243–8.

Hays, Samuel P. *Conservation and the Gospel of Efficiency: The Progressive Conservation Movement, 1890–1920*. Cambridge, MA: Harvard University Press, 1959.

Hedrén, Johan and Bjorn-Ola Linnér. 'Utopian Thought and the Politics of Sustainable Development'. *Futures* 41, no. 4, (2009): 210–19.

Heller, Michael A. 'The Tragedy of the Anticommons: Property in the Transition from Marx to Markets'. *Harvard Law Review* 111, no. 3 (1998): 621–88.

Hess, Charlotte. 'Mapping the New Commons'. Presented at 'Governing Shared Resources: Connecting Local Experience to Global Challenges', the twelfth Biennial Conference of the International Association for the Study of the Commons, University of Gloucestershire, Cheltenham, 14–18 July 2008.

Hooper, Virginia and Bruce A. Lankford. 'Unintended Water Allocation: Gaining Share from the Ungoverned Spaces of Land and Water Transformations', in Ken Conca and Erika Weinthal (eds), *Oxford Handbook of Water Politics and Policy*. Oxford: Oxford University Press, forthcoming, 2016.

Keys, Patrick, Jennie Barron, and Mats Lannerstad. *Releasing the Pressure: Water Resource Efficiencies and Gains for Ecosystem Services*. Nairobi: United Nations Environment Programme, Stockholm Environment Institute, Stockholm, 2012.

Lankford, Bruce. *Resource Efficiency Complexity and the Commons: The Paracommons and Paradoxes of Natural Resource Losses, Wastes and Wastages*. Abingdon: Routledge, 2013.

Lawrence, Mark. 'Heartlands or Neglected Geographies? Liminality, Power, and the Hyperreal Rural'. *Journal of Rural Studies* 13, no. 1 (1997): 1–17.

Leach, Melissa, James Fairhead, and James Fraser. 'Green Grabs and Biochar: Revaluing African Soils and Farming in the New Carbon Economy'. *Journal of Peasant Studies* 39, no. 2 (2012): 285–307.

Lopez-Gunn, Elena, Pedro Zorrilla-Miras, Fernando Prieto, and M. Ramon Llamas. 'Lost in Translation? Water Efficiency in Spanish Agriculture'. *Agricultural Water Management* 108 (2012): 83–95.

Luque, Rafael and James H. Clark. 'Valorisation of Food Residues: Waste to Wealth Using Green Chemical Technologies'. *Sustainable Chemical Processes* 1, no. 10 (2013).

Manson, Steven M. 'Simplifying Complexity: A Review of Complexity Theory'. *Geoforum* 32, no. 3 (2001): 405–14.

Manson, Steven. M. 'Does Scale Exist? An Epistemological Scale Continuum for Complex Human-Environment Systems.' *Geoforum* 39, no. 2 (2008): 776–88.

MEA. *Ecosystems and Human Well-being: Synthesis*. Washington, DC: Millennium Ecosystem Assessment, 2005, at: www.millenniumassessment.org/documents/document.356. aspx.pdf (accessed 18 January 2016).

Murray-Rust, D. Hammond, and W. Bart. Snellen. *Irrigation System Performance Assessment and Diagnosis*. Sri Lanka: International Irrigation Management Institute, 1993.

Neuman, Janet. 'Beneficial Use, Waste, and Forfeiture: The Inefficient Search for Efficiency in Western Water Use'. *Environmental Law* 28 (1998): 919–96.

Norgaard, Richard B. 'Ecosystem Services: From Eye-Opening Metaphor to Complexity Blinder'. *Ecological Economics* 69, no. 6 (2010): 1219–27.

Norris, Joe. 'Montana V. Wyoming: Is Water Conservation Drowning the Yellowstone River Compact?' *University of Denver Water Law Review* 15, no. 189 (2011).

OECD. *Green Growth Studies: Energy*. Paris: The Organisation for Economic Co-operation and Development, 2011.

Polimeni, John M., Kozo Mayumi, Mario Giampietro, and Blake Alcott. *The Jevons Paradox and the Myth of Resource Efficiency Improvements*. London: Earthscan, 2008.

Raymond, Eric. 'The Magic Cauldron'. *ALS'99: Proceedings of the 3rd annual conference on Atlanta Linux Showcase – Volume 3*. Berkeley, CA: USENIX Association, 2000, at: www.catb.org/esr/writings/cathedral-bazaar/magic-cauldron/ (accessed 18 January 2016).

Reijnders, Lucas. 'The Factor "X" Debate: Setting Targets for Eco-Efficiency'. *Journal of Industrial Ecology* 2, no. 1 (1998): 13–22.

Ruth, Matthias, Eugenia Kalnay, Ning Zeng, Rachel S. Franklin, Jorge Rivas, and Fernando Miralles-Wilhelm. 'Sustainable Prosperity and Societal Transitions: Long-Term Modeling for Anticipatory Management'. *Environmental Innovation and Societal Transitions* 1, no. 1 (2011): 160–5.

Sax, Joseph L. 'The Constitution, Property Rights and the Future of Water Law'. *University of Colorado Law Review* 61 (1990): 257–82.

Schmidneiny, Stephan, and Björn Stigson. *Eco-Efficiency: Creating More Value with Less Impact*. World Business Council for Sustainable Development, 2000, at: www.wbcsd.org/web/publications/eco_efficiency_creating_more_value.pdf (accessed 18 January 2016).

Scoones, Ian. 'New Ecology and the Social Sciences: What Prospects for a Fruitful Engagement?' *Annual Review of Anthropology* 28, no. 1 (1999): 479–507.

Shupe, Steven J. 'Waste in Western Water Law: A Blueprint for Change'. *Oregon Law Review* 61 (1982): 83–499.

Smith, Henry. 'Governing Water: The Semicommons of Fluid Property Rights'. *Arizona Law Review* 50, no. 2 (2008): 445–78.

Socolow, R., C. Andrews, F. Berkhout, and V. Thomas. *Industrial Ecology and Global Change*. Cambridge: Cambridge University Press, 1996.

Strathern, Marilyn. 'Social Relations and the Idea of Externality', in Colin Renfrew and Chris Scarre (eds), *Cognition and Material Culture: The Archaeology of Symbolic Storage*. Cambridge: McDonald Institute for Archaeological Research, 1998, 135–47.

Strathern, Marilyn. 'Environments Within: An Ethnographic Commentary on Scale', In Kate Flint and Howard Morphy (eds), *Culture, Landscape, and the Environment. The Linacre Lectures 1997*. Oxford: Oxford University Press, 2000, 44–71.

Tainter, Joseph A. 'Energy, Complexity, and Sustainability: A Historical Perspective'. *Environmental Innovation and Societal Transitions* 1, no. 1 (2011): 89–95.

Underdal, Arlid. 'Complexity and Challenges of Long-Term Environmental Governance'. *Global Environmental Change* 20, no. 3 (2010): 386–93.

Van Gennep, Arnold. *The Rites of Passage*. Translated by Monika B. Vizedom and Gabrielle L. Caffee. London: Routledge and Kegan Paul, 1960 [1909].

Ward, Frank A. and Manuel Pulido-Velázquez. 'Water Conservation in Irrigation Can Increase Water Use'. *Proceedings of the National Academy of Sciences* 105, no. 47 (2008): 18215–20.

Warner, Rosalind. 'Ecological Modernization Theory: Towards a Critical Ecopolitics of Change?' *Environmental Politics* 19, no. 4 (2010): 538–56.

Yanlk, Lerna K. 'Constructing Turkish "Exceptionalism": Discourses of Liminality and Hybridity in Post-Cold War Turkish Foreign Policy'. *Political Geography* 30, no. 2 (2011): 80–9.

York, Richard and Eugene A. Rosa. 'Key Challenges to Ecological Modernization Theory: Institutional Efficacy, Case Study Evidence, Units of Analysis, and the Pace of Eco-Efficiency'. *Organization & Environment* 16, no. 3 (2003): 273–88.

6 The right to not be excluded

Common property and the struggle to stay put[1]

Nicholas Blomley

shout here we are
amazingly alive
against all long odds
left for dead
shoutin this death culture
dancin this death culture
out of our heads
amazingly alive
Bud Osborn, 'Amazingly Alive',
                (Osborn 1999: 10)

Introduction

Private property, and the right to exclude is territorialised through the multiple 'zones of exclusion' that hedge out the urban poor, such as residents facing the displacement generated by inner city gentrification. Any stake that such residents may have, if acknowledged at all, is rarely seen in terms of a right, let alone a common property right. On what basis might such a right rest, and how might we think of it in terms of property?

Residents in Vancouver's Downtown Eastside make powerful claims against exclusion, and for inclusion. C.B. Macpherson defines common property as the right to not be excluded. My goal here is to place the two in conversation, drawing the lessons of both to enrich the theory of commons, and the praxis of urban struggle. Macpherson points us to the exclusionary logics of private property in relation to a common property right to not be excluded. He argues for an enriched reading of human liberty that should be served by property, including common property. In so doing, he reminds us that the experiential force of exclusion and the salience of the right to not be excluded are socially differentiated. His recognition of common property as a relation also points to the need to attend to the practice of commoning, avoiding the ethical and analytical dangers that arise when we posit 'the commons', with its resultant boundedness, fixity, and zero-sum membership.

Such claims, I suggest, help us in thinking of the struggle in a place such as the Downtown Eastside as the exercise of a *property* right. While operating always under the shadow of powerful forms of capitalist exclusion, the right to not be excluded sustains pluralistic forms of human capacity and potentiality, through multiple forms of practice and creative enactment. Treating commons as a right, rather than a set of resources, or an autonomous space, also allows us to recognise the multiple and historically layered struggles for inclusion, and against exclusion, in urban space. In grounding Macpherson's abstract claims, and situating them in a site of struggle, we can learn from the margins.

The right to stay put

Staggering numbers of people within the Western world are forced from their homes. In the United States, for example, some 1.8 million households are displaced annually (i.e. approximately five million people in total). In general terms, we can observe two things: these tend to be marginalised people, with more insecure forms of tenure, and the proximate cause of most displacement is private-sector, rather than public-sector action.[2] Displacement caused by gentrification is a particular concern, as it is likely to disproportionately affect the most vulnerable people. Vancouver's marginalised Downtown Eastside neighbourhood is one such example.[3] Growing development pressures threaten the long-term survival of this precious and vulnerable community. For over a generation, residents within the Downtown Eastside have bravely and insistently argued for the 'right to stay put' (Hartman 2002; cf. Imbroscio 2004, Maeckelbergh 2012), predicated on the claim that:[4]

> *[m]oving people involuntarily from their homes or neighbourhoods is wrong.* Regardless of whether it results from government or private market action, forced displacement is characteristically a case of people without the economic and political power to resist being pushed out by people with greater resources and power, people who think they have a 'better' use for a certain building, piece of land, or neighborhood. The pushers benefit. The pushees do not.
>
> (Hartman *et al.* 1982: 4–5, original emphasis)

'Pushing' has produced and shaped that which is now the Downtown Eastside for a 150 years. This gridded terrain of blocks and lots was carved out from indigenous lands, imagined as *terra nullius*, to be used by a settler society that continue to imagine it has a better use for it. In a bitter irony, a significant number of the contemporary residents of the Downtown Eastside are indigenous. Driven from their lands by the expulsions of settler capitalism, many are forced into the urban margins. Japanese-Canadians resident in the Powell Street Nihonmachi neighbourhood were also forcibly 'pushed' from their homes during the Second World War, with their possessions liquidated by the state. African-Canadian residents were expelled from their homes in Hogan's Alley to make way for a viaduct in the early 1970s, as were residential hotel occupants during the preparation for Vancouver's Expo World's Fair in 1986. These historical layers now combine with the current round of 'pushing', as developers look to the Downtown Eastside as the next 'frontier'. Echoing the colonial logics that created the space, a contemporary condo marketer invites urban 'pioneers' to colonise 'undiscovered territory'.[5]

But the 'pushees' have long avowed their 'right to stay put', as evidenced in two recent related initiatives in the Downtown Eastside. *Zones of Exclusion* was a report of the grass-roots Downtown Eastside Neighbourhood Council Action Committee, released in early 2011 to coincide with a struggle over a proposal to increase density in the Chinatown neighbourhood of the Downtown Eastside.[6] The report documents a series of 'zones of exclusion', including new condos, up-scale shops and restaurants, noting their up-scale branding (condos replete with 'terrazzo stone slab countertops', 'unique elevator car storage', 'clear-bottom acrylic rooftop', '12-metre pool', and so on); their implication in logics of exclusion and surveillance (security cameras, economic screening, and differential pricing); and their erasure and conversion of a place of struggle, dignity, and marginalisation into a rebranded ersatz playground of 'heritage' and hipsterdom, including the redevelopment of the Wing Sang building, the oldest in Chinatown, by Bob Rennie (he who invited urban pioneers to settle the urban frontier). Rennie is known locally as the 'Condo King' for his formidable ability to market and sell Vancouver's real estate lifeblood. Rennie Realty is quoted as promising to disclose the complete history of a property, so as to protect clients from fraud. 'One wonders if "disclosing the complete history of a property"', dryly questions the Report's authors, 'includes the history of colonization, and the theft of indigenous land, or the displacement of low-income residents from their community?' (24).

The report notes that gentrification not only ushers in increasing land values and higher rents, but also 'produces a kind of internal displacement for low-income residents by creating zones of exclusion' (3). These zones include spaces that poor people are excluded from because they lack the economic means, or from which they are expelled by virtue of intensified policing. But exclusion is at work in a more generalised sense. As land is used to build housing for the rich, it no longer becomes available to the low-income community. In this sense, it is said, 'gentrification excludes possibilities' (3). Ultimately, 'as gentrification produces more and more zones of exclusion, low-income residents become alienated from their own community. It is the experience of internal displacement – the feeling of being out of place in one's own neighborhood' (3).

The report is far from dispassionate. It is wry, sardonic, angry, and powerfully affirmative: 'Without a fight, everything will be lost. But as history reveals, Chinatown is a community where resistance won't go away' (12). Chinatown is characterised as an historic site (a 'hand-built community'), a haven for low-income residents, and a centre for community engagement, resistance against racism, and for civil rights struggles. Residents are described as successfully resisting racist vigilantes in the early race riots, whereby 'the Chinese community strengthened, rebuilt itself and asserted their right to remain'.

Exclusion not only constrains, but also takes something away. A sense of what is at stake is evident in a second initiative, but one that emphasises inclusion. This is the attempt to carve out a 'social justice zone' in the core of the Downtown Eastside (Figure 6.1). Such a zone protects principles and resources that are of value. It is portrayed as a space where low-income people and their basic human and social needs 'have priority over profit'. Necessarily, it is imagined as 'a place where low income and vulnerable people have a right to be and won't be pushed out.' The creators of the social justice zone ask: 'what could happen here that would respect the basic human rights of low income people to chose where they want to live and to have basic needs met?'.[7]

Struggles relating to gentrification are, of course, a familiar story, repeated throughout cities across the world. Curiously, however, when thinking about Western cities, the workings of property as a bundle of relations between people, predicated on rights and interests, do not seem to loom very large in our accounts. Elsewhere (Blomley 2004), I have tried to reflect on this, noting the tendency to view Western urban spaces as 'settled' in relation to property. Such a view, however, belies the often turbulent and violent processes by which property relations are remade as urbanisation unfolds – sharply evident in a settler city such as Vancouver – as well as the evident diversity of property forms and relations, both formal and unacknowledged, that can be discerned when one starts to look: 'A process of displacement and dispossession, in short... lies at the heart of the urban process under capitalism' (Harvey 2012: 18).

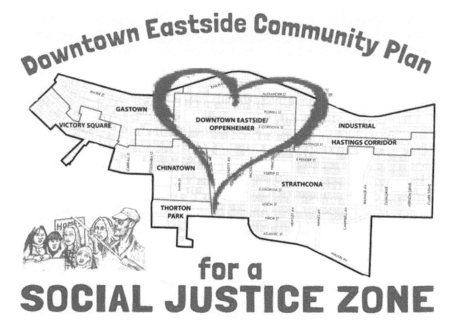

Figure 6.1 The social justice zone

Source: Reproduced by kind permission of DTES resident and activist Diane Wood.

The complex and interlocking nature of such displacements in Vancouver's Downtown Eastside is evident in the comments of a Nisga'a man called Gwin ga'adihl amaa goot, also known as Herb Varley, who rose before City Council to oppose the proposal for increased density. A resident in the area, and the President of the Board of Directors for the Downtown Eastside Neighbourhood Council that produced the *Zones of Exclusion* report, his comments offer a rich account of displacement, silencing, and marginalisation. He describes the challenges of minimal welfare benefits and rising rents, as well as the social violences caused by gentrification. He also combines this with the recognition of the Downtown Eastside as the traditional lands of indigenous peoples, as well as characterising the area as being like an Indian reserve. By this I assume he means both that it contains and corrals many indigenous people, and also, as he puts it, 'people got put there because it wasn't deemed valuable enough'. Yet now, he complains, as indigenous people are 'getting kicked out because they want our resource', so the Downtown Eastside has become a valued resource, and the residents face being 'kicked out' again. 'I hope you were listening,' he asks Council, 'I hope you were listening'.[8]

The City wasn't listening. So we must. Why is it wrong for Herb Varley and other residents to be kicked out? If Downtown Eastside residents have a right to stay put, as Chester Hartman puts it, on what is this right based? How might we grapple with the multiple layers of displacement at work in a place such as the Downtown Eastside, including that which made the place to begin with? How might we think of his resistance both in terms of the exclusion Gwin ga'adihl amaa goot opposes, and the valued potentiality that he protects?

It is useful, as a starting point, to think of the zones of exclusion and social justice together. The reference to 'exclusion' points us to a set of power relations that deny access. And exclusion is only significant if there is something valued in the zones (material and metaphorical) from which one has been locked out. At work here, I want to suggest, is both an argument *against* exclusion, and *for* continued access to something of value (Blomley 2008). This is not simply a negative argument against exclusion, in other words, but also a positive claim. The latter entails some sort of interest or claim in the Downtown Eastside on the part of the urban poor, both as individuals and as a collective. As echoed in the social justice zone, this not only includes material assets, such as shelter, but also entails a valued way of life and relations. This is not a formally sanctioned claim, but is nevertheless performed through and sustained by dense historical geographies of use, domicile, collective resistance, and inherent need (Blomley 2008). It is not necessarily an exclusive claim, but rather one that insists on the inclusion of poor residents in the making and remaking of urban space: the social justice zone is 'a place where low income residents are recognized as the experts in matters that affect them and have control over decisions, services and operations that affect them'.[9]

It is crucial to try and think of these claims against exclusion and for inclusion simultaneously. One obvious solution is to appeal to the commons, as well as a concern at 'enclosure'. Were we to think of the Downtown Eastside through this lens, we can do so in a number of ways. One option is to draw from the common-pool resource literature, with its emphasis on rational choice (e.g. Garnett 2012; Lee and Webster 2006). However, the methodological individualism and instrumental orientation of this literature does not seem to lend itself to the Downtown Eastside. This is not an Ostrom-like 'common pool regime', with an emphasis on choice, coordinated resource use and formal governance. Rather, it is an angry, moral, affective, and political set of claims, which deploy a language of rights and justice. A second alternative is to draw from contemporary progressive accounts that see the commons as a space of collectivity and cooperation, horizontal organisation and 'emancipation through convivial connections' (Helfrich 2012: 36). However, while more normatively appealing, I find much of this literature to be both analytically vague and normatively un-nuanced. How, for example, are we to distinguish, say, the 'commons of the rich' (Blackmar 2006), such as a condo or gated community, from the claims of the Downtown Eastside?

The right to not be excluded

Motivated by the exploratory spirit of this volume, I thus turn to the curiously neglected work of C.B. Macpherson.[10] A political economist, trained at the London School of Economics with Harold Laski, where he met Richard Henry Tawney and Morris Ginsberg, his writing is animated by a focus on the ethical and political dimensions of property. 'All roads lead to property' (Macpherson 2013: 121), he insisted. But his route map, as we shall see, was unconventional. Unusually, he moved consciously and capably between political economy (he was once described as 'five-sixths of a Marxist': see Svacek in Carter 2005: 832), and a liberalism committed to equality and democracy.

Macpherson offers a brief, but productive definition of common property, which he consciously sets in distinction to private property:

> *common property is created by the guarantee to each individual that he will not be excluded from the use or benefit of a thing*; private property is created by the guarantee that an individual can exclude others from the use or benefit of something.
>
> (Macpherson 1978: 5, my emphasis)

There is, as we shall see, a lot riding on the coat-tails of this claim. And there are things that are less useful, or need supplementing. However, I think that there may also be some valuable insights and nudges at work here. This chapter, therefore, is an attempt to use Macpherson to think about common property, with particular reference to the Downtown Eastside (and, hopefully, in relation to gentrification-induced displacement more generally). In so doing, we can ground Macpherson's abstract account, and productively enrich it, drawing upon the lived experience of struggle.

Some prefatory comments: Macpherson's deliberately refutes the dangerous misconception that property is identical with private property. For historical reasons, he argues, 'we have all been misled by accepting an unnecessarily narrow concept of property' (1978: 201). Property should not be confined to the right to exclude, he insists, 'but may equally be an individual right not to be excluded by others' (Macpherson 1978: 201). If we accept that property is an individual right, not a thing, created and enforced by the state, there is no logical reason to require that property be confined to private property, he argues. The justification of property as derivable from an individual's essence, a 'right to some use or benefit of something without the use or benefit of which he [*sic*] could not be fully human' (Macpherson 1978: 202), is not logically entailed in exclusivity.

Note further that Macpherson treats property as a right of individual persons. While this is rather provocative, particularly when it comes to common property, it is a deliberate choice. As noted below, Macpherson's justification for property rests on its value in achieving individual human ends, which he treats as 'the most solid basis on which to justify an institution or a right' (Macpherson 1978: 201). However, while attentive to individual liberties, his trenchant critique of possessive individualism (Macpherson 1962) should remind us that he is no individualist. Individuals, he notes, develop their potential in relation to others. Moreover, property is also fully social and political, rather than some pre-natural essence. It is a human institution, which requires a collective body to enforce it, policing a system of inter-personal rights.[11]

Strategic extension

At the outset, it is important to address the merits of thinking about the struggle to 'stay put' in a place like the Downtown Eastside through a property lens at all. Why not rely upon a more generic language of *rights* (such as the 'right to the city') or, given some of the anxieties relating to rights-talk more generally, simply use a language of *need*? Some may also resist the idea of characterising this as a form of *property* interest (or even property right) at all, given a reasonable concern as to the individualised manner in which property has been taken up within liberal-capitalism. However, as we shall see, Macpherson asks us to recognise the inherent diversity and potential possibility of property, understood as a relation between people in regard to some valued resource.

Indeed, to be able to characterise such struggles as expressions of common right seems to me strategically useful in a number of senses. Most immediately, it allows us the chance to talk back against property theory, which tends to adopt an impoverished view of property, confining it to private or occasionally state forms, with an narrow focus on property's exclusionary dimensions (Blomley 2004). Given the discursive ubiquity of commoning, it also becomes strategically useful to hook up a local struggle to a global movement. The ability to do so provides a powerful discursive resource. Further, it allows us to politicise an urban conflict that tends to be framed as a matter of land use, premised on planning's question, 'where do things belong?' As Krueckeberg (1995: 301) notes, the effect is to efface a more pressing question: to whom do things belong?

'Where things belong,' he argues, 'cannot be answered justly until we know whose things we are talking about.'

Macpherson, I suspect, would agree with such a framing, arguing for the merit of using property (albeit in a richer form) to articulate broader claims relating to democracy and a right to a kind of society, suggesting that these claims:

> will not be firmly anchored unless they are seen as property. For, in the liberal ethos which prevails in our liberal-democratic societies, property has more prestige than almost anything else. And if the new claims are not brought under the head of property, the narrow idea of property will be used, with all the prestige of property, to combat them.
>
> (Macpherson 2012: 138)

The danger, of course, is that we fall back to 'the narrow idea of property', understood purely as *private* rights to things, justified in instrumental terms.[12] While Macpherson draws from the rich history of property to point us to its protean and broad meanings (cf. Pierson 2013), the tendency has been to focus debate on private property. This is to be expected, he notes, given that exclusivity is a contentious moral issue. As private property took on a central role in the allocation of labour and resources in the emergence of capitalism, it became treated as synonymous with property as a whole. Yet this is a mistake: 'We have treated as the very paradigm of property what is really only a special case,' Macpherson (1978: 201) insists. Property is thus a far more commodious category than we might suppose. While it is tempting for the anti-displacement activist and Leftish academic to inveigh against 'property', Macpherson's point is that we need to be more careful with how we treat such a category, his primary goal being to 'restore to the term the complexity of which modern Western societies had stripped it' (Lindsay 1996: 123). In this he was likely influenced by Tawney's argument that:

> property is the most ambiguous of categories. It covers a multitude of rights which have nothing in common except that they are exercised by persons and enforced by the state.... It is idle, therefore, to present a case for or against private property without specifying the particular forms of property to which reference is made.... The course of wisdom is neither to attack private property [nor, I would add, common property] in general nor to defend it in general.... It is to discriminate between the various concrete embodiments of what, in itself, is, after all, little more than an abstraction.
>
> (Quoted in Lindsay 1996: 123)

Exclusion and inclusion

How, then, might we think about property in its 'concrete embodiments' within the Downtown Eastside, and how might this help us in theorising common property in particular? Most immediately, Macpherson's account pushes us to think relationally. The 'right to not be excluded' is inherently both a negative and a positive right: it is quite deliberately not a right 'to be included'. Exclusion is problematic precisely as it negates access. This is useful in directing us to the importance of recognising the relation between common right and the logics of exclusion central to private property. Both need to be thought against and with each other, as we can see most sharply in a marginalised place such as the Downtown Eastside (cf. Jeffrey *et al.* 2012; Lang 2014). This becomes essential in understanding the historically layered, yet enduring, struggles against displacement and dispossession in the neighbourhood.

This is useful, given a tendency to overlook such exclusionary logics, characterising it purely as an autonomous space, set beyond such relations. This is evident in the title of a recent compendium on common property: 'The wealth of the commons: a world *beyond market and state*' (Bollier and Helfrich 2012, my emphasis). As one contributor notes, the commons can be thought of as 'the shared resources which people manage by negotiating their own rules through social or customary traditions, norms and practices' (Quilligan 2012: 72). While activists in the Downtown Eastside may strive for an autonomous zone of social justice, it is clear here that common right and private right cannot be so easily separated. This applies not only through the process whereby 'zones of exclusion' are imposed upon a community, but is also evident in collective claims made over resources formally held by private or state actors (Blomley 2008; Lang 2014). Common right, at least within Vancouver's Downtown Eastside, is not a 'world beyond market and state' (Bollier and Helfrich 2012) but deeply entangled in exclusionary property relations.

Downtown Eastside activists also know full well that such relations are not open-ended, but materialised in systematic ways, producing a propertied gradient, predicated on and productive of class domination. Similarly, as a political economist, Macpherson situates the right to exclude, and the right to not be excluded, within a capitalist market economy. It is this that gives the right to not be excluded a particular salience. Exclusion, Macpherson (1962) alerts us, is structured by a set of powerful beliefs that he terms possessive individualism, based on his reading of Hobbes and Locke:

> Denial or limitation of access is a means of maintaining class-divided societies, with a class domination which thwarts the humanity of the subordinate and perverts that of the dominant class; this is a condition which neither any amount of 'consumers' sovereignty', nor the fairest system of distributive justice can offset or remedy.... The extent and distribution of that access is set by the system of property.
>
> (Macpherson 2012: 120)

This, to me, seems an important point, all too evident in the Downtown Eastside, particularly when the gendered and racialised dimensions of capitalist exclusion are also recognised. City space is 'saturated space', as Huron (2015) puts it. Exclusion is productive of and produced through particular social configurations that generate systematic and deeply entrenched relations. The experiential force of exclusion and the salience of the right to not be excluded are socially differentiated (Purser 2014). If we attend only to the internal rules through which 'the commons' is constructed, we are danger of missing this crucial social gradient. The iconic middle-class gated community operates, for instance, through the supposed 'core beliefs' of the commons, such as cooperation, non-rivalry, inter-relationality, horizontality, and 'emancipation through convivial connections' (Helfrich 2012: 36). Yet its relationship to capitalist property relations and markets is radically different than commoning in Vancouver's inner city, and needs to be analytically and ethically distinguished in ways that many scholarly claims to the commons fail to register. Commons are not all of a piece, as Amin and Howell remind us, in their introduction to this volume, and are frequently liable to be hijacked by elite interests (Blackmar 2006). This becomes essential when we grapple with the multiple forms of displacement in a setting such as the Downtown Eastside, including its foundational colonial moment. A careful attention to the workings of capitalism, and the manner in which land is commodified, is central to an understanding of these multiple forms of displacement, refracted through the hierarchical relations of property.

From nouns to verbs

One of Macpherson's main claims is that property is not a thing, but a right. Property, he argues, is, at base, a 'man-made institution which creates and maintains certain relations between people' (Macpherson 1978: 1). In this, he echoes Harvey's view that:

> the common is not to be construed … as a particular kind of thing, asset or even social process, but as an unstable and malleable social relation between a particular self-defined social group and those aspects of its actually existing or yet-to-be-created social and/or physical environment deemed crucial to its life and livelihood.
>
> (Harvey 2012: 73)

While things are clearly crucial to property, the recognition of its relational dimensions seems useful. We should avoid the tendency to noun the verb, and to think of *the* commons, considered as a singular thing. We do not speak of private property as an object, yet rely upon a propositional language that asks us to nail down the essence of 'the' commons, and in so doing, to invite us to think of it as prior to its enactments and relationships (White 1990).

Macpherson's attention to exclusion/inclusion, however, takes us towards a more active, verb-oriented sense of *commoning*, where we 'think first not of title deeds, but of human deeds' (Linebaugh 2008: 45). Property, of whatever form, should be thought of as complex, more-or-less successful performances that are enacted, embodied, and practised (Blomley 2013). 'The commons' does not exist prior to such performances. Commoning in the Downtown Eastside is enacted through acts such as protest, anti-gentrification walking tours, poetry, occupations and squats, and everyday acts of presence, reiterated use, and engrained inhabitation. Such performances more or less successfully produce the 'effect' that is common property (and indeed its 'affect' too). They have an iterative quality, being called upon as a political resource, demonstrating collective agency and capacity (cf. Bresnihan and Byrne 2015; and McFarlane and Desai: Chapter 9).[13]

Grounded in a particular and intensely occupied place, commoning produces physical and representational landscapes, as I have noted elsewhere (Blomley 1998, 2004, 2008; Jeffrey: Chapter 7). But while commoning is intrinsically spatial (and temporal) we need to be cautious of appealing to 'the' commons as a finite space. In part, I think, this may reflect a nostalgia for the classic English rural commons (and one, moreover, that neglects the fact that it was much more than a fixed space, but reliant on networks of access and use over land held by multiple interests).[14] Commoning may generate spatial distinctions (such as the social justice zone, noted above), but such spaces must be thought in relational-strategic, rather than fixed terms. To read commons through a fixed spatiality is to invite a zero-sum politics that negates the necessary connections that exist between the multiple commoners in a place such as the Downtown Eastside. Similarly, the territory of private property is best thought of not in terms of a fixed geography, but as a device for the arrangement of the relations that constitute it (Blomley 2015).

Liberty and inclusion

As noted, Macpherson characterises common property as a *right* (not a claim, or moral imperative) to not be excluded. This is centrally important, as rights shift relationships out of the realm of the desirable and into the domain of the essential. To endorse a rights-designation is to approve a distribution of freedom and authority, and a particular view of 'what may, must, and must not be done' (Wenar 2011: n.p.). To characterise property – whether private or common – as a right, Macpherson notes, requires justification: 'Property is not thought to be a right because it is an enforceable claim: it is an enforceable claim because it is thought to be a human right' (1978: 11). How then is property a human right? Here we should visit another side to Macpherson. As noted, while he is a political economist, he is also a (strategic) liberal. The productive tension between liberal values of individual freedom and 'the deeply rooted inequalities in the institutional order which are shielded by the structure of property ownership ... forms the underlying unity of Macpherson's thought' (Leiss 1988: 15). Macpherson claimed that his central goal was to:

> work out a revision of liberal-democratic theory, a revision that clearly owed a great deal to Marx, in the hope of making that theory more democratic while rescuing that valuable part of the liberal tradition which is submerged when liberalism is identified as synonymous with capitalist market relations.
>
> (Macpherson 1976: 423)

As a liberal, Macpherson places great weight on the affirmation of individual liberty (cf. Purdy 2005). Property rights, liberalism tells us, are valuable to the extent that they maximise human liberty. But the ends that liberty sustains can be defined in at least two ways, Macpherson notes. On one dominant view, humans are first and foremost the appropriators of material utilities. From this perspective, property is valuable to the extent that it maximises individual satisfaction or utility. A second conception, however, that has fallen from view, but one which Macpherson seeks to revitalise, is based on a view of humans not as consumers, but as doers and creators, exercising uniquely human powers. The ability of human beings to exercise their own capacities is a marker of essential human dignity, such that: '[a] good life is one which maximizes these powers. A good society is one which maximizes (or permits or facilitates the maximization of) these powers, and thus enables men [*sic*] to make the best of themselves' (Macpherson 1973: 8–9). There is a clear link here to Aristotelian virtue ethics (Nussbaum 1997) and their uptake within 'progressive property theory' (Alexander and Peñalver 2009), although supplemented with a refreshing materialism.

The right to not be excluded, therefore, can be said to be valuable in that it provides access to the resources necessary for people to advance their 'uniquely human capacities'. These non-utilitarian ends to which property can be put, Macpherson notes, are pluralist, diverse, and often incommensurable.[15] A rich vein of poetry and art in the Downtown Eastside similarly attests to these diverse ends, including tolerance, acceptance, joy, and creativity. Such individual ends may become collective values, including those that affirm and sustain others. As the Downtown Eastside street poet Bud Osborn puts it: 'the power of love holds us together/not as passive abstraction or a commodity privatized/but love/as fiery personal and collective social justice passion...' (Osborn 1998: 285)

Viewed carefully, such a perspective offers useful tools in tackling a difficult ethical question. The comments of Gwin ga'adihl amaa goot noted above point us not only to the historical layers of colonial exclusion that are the very condition for the Downtown Eastside, but also to the continued presence of such colonial logics. Dave Diewert, one of the authors of the *Zones of Exclusion* report, responded to my argument by worrying that an appeal to the commons is in danger of erasing such historically layered displacements. However, a view of common property as a right to not be excluded comprising multiple rights-bearers who are both recipients and trustees, as opposed to a more static and narrowly bounded conception of a 'commons' made up of a fixed and unitary membership, seems more useful here. It is not groupness that grounds commoning, suggests Macpherson, but its non-exclusive relationality. Commoning need not be premised on an essentialised singularity, particularly one based on fixed notions of community and territory, but rather on a properly *political* relationship, as Amin and Howell suggest in their introduction to this volume, structured according to an ethic of obligation. Moreover, as noted above, a pluralist understanding of the ends of property, based on human capacities, allows us to affirm the essential connections between the right to not be excluded of the Nisga'a,

Cree, or Somali refugee, living on welfare in the Downtown Eastside, the white sex-trade worker struggling to survive, the working-class Japanese-Canadian owner expelled from Powell Street in 1942, and the Musqueam, Squamish, and Tsleil-Waututh Nations, for whom the Downtown Eastside remains traditional territory.

A broader conception of property

Bud Osborn's poetry of the Downtown Eastside affirms the very survival of the poor, enduring in one of the hottest real estate markets in North America: 'shout here we are,' he declares, 'amazingly alive/against all long odds/left for dead/ shoutin this death culture/dancin this death culture/out of our heads/amazingly alive' (Osborn 1999: 10).

But as Bud's poem notes: 'here we are/against long odds'. The deck is stacked, of course, because the right to exclude is privileged under capitalism. As Macpherson puts it, the centrality of individual utility clashes with the capacity of individuals to develop their human potential, given the workings of a capitalist property system:

> [W]hen the liberal property right is written into law as an individual right to the exclusive use and disposal of parcels of the resources provided by nature and of parcels of the capital created by past work on them, and when it is combined with the liberal system of market incentives and rights of free contract, it leads to and supports a concentration of ownership and a system of power relations between individuals and classes which negates the ethical goal of free and independent individual development.
>
> (Macpherson 1978: 200)

Freedom for the pike, Macpherson's mentor Tawney noted, is often death to the minnow. The centrality of the right to exclude in a place such as the Downtown Eastside negates human fulfilment. Displacement not only compromises those higher ends, but also threatens annihilation:

> Eviction from the neighbourhood in which one was at home can be almost as disruptive of the meaning of life as the loss of a crucial relationship. Dispossession threatens the whole structure of attachments through which purposes are embodied, because these attachments cannot readily be re-established in an alien setting.
>
> (Marris 1986: 57; cf. Strabowski 2014)

Liberty necessarily includes not only one's natural capacities (energy, skill) but also one's ability to exert them. It therefore includes *access*. 'It must therefore treat as a diminution of a man's [*sic*] powers whatever stands in the way of his realizing his human end, including any limitation of that access' (Macpherson 1973: 9). This leads us to a central paradox:

If, as liberal theory asserts, an individual property right is required by the very necessities of man's [*sic*] nature and condition, it ought not to be infringed or denied. But unless it is seriously infringed or denied, it leads to an effective denial of the equal possibility of individual human fulfillment.

(Macpherson 1978: 200)

Liberalism, for Macpherson, will continue to be contradictory until it recognises and resolves this paradox. Without this, it merely legitimates inequality. Of necessity, he argues, property must not be confined to a right to exclude others, but must equally be recognised as an individual right not to be excluded. As such, property is both a constraint upon liberty, *and* its very precondition.[16] However, this requires a reassessment of what we understand by property, and the ends to which it works. Macpherson thus engages in a productive form of categorical politics, seeking to make property more capacious, and pushing back at the tendency to confine property to *private* property (cf. Gibson-Graham, Cameron, and Healy: Chapter 12). The liberal solution, Macpherson notes, has been to focus on narrowing an individual property right. We are starting in the wrong place, he insists. The property right of liberal theory is too narrow, treating property as exclusively *private* property. Property need not be confined to a right to *exclude* others, but must also entail an individual right *not* to be excluded:

When property is so understood, the problem of liberal-democratic theory is no longer a problem of putting limits on the property right, but of supplementing the individual right to exclude others by the individual right not to be excluded by others.

(Macpherson 1973: 201)

Such lessons are broader still. Macpherson extends his argument by directing us to a broader conception of democracy predicated on a developmental ethic:

If property is to remain justified as instrumental to a full life, it will have to become the right not to be excluded from the means of such a life. Property in such circumstances will increasingly have to become a right to a *set of social relations, a right to a kind of society*.

(Macpherson 2013: 138, my emphasis)

As any adequate democratic theory must treat democracy as a kind of society and must regard the individual members as at least potentially doers rather than mere consumers, it 'must assert an equal effective right of the members to use and develop their capacities,' he argues (Macpherson 2013: 51; cf. Singer 2009).

Macpherson points to the labour unionisation and welfare capitalism of his day as marking such an optimistic change in the meaning of property. We have, in our own neoliberal era, to go elsewhere. I would suggest that we can draw from the 'margins', such as the Downtown Eastside, to provide a richer account of property's political potential, particularly in relation to common property. Van der Walt (2010; see also Peñalver and Katyal 2007) argues against a logic of centrality, which privileges private property, and treats the marginal owner as weak, vulnerable, and dependent. Traditionally, he notes, the law knows only two ways to deal with the marginal: prescribe rules that govern their presumptively unlawful interactions with the property rights of the privileged; and ignore them completely for being marginal, and hence irrelevant to the development of property's laws and principles. Both such strategies are painfully evident in the Downtown Eastside. But it is precisely in the margins that law becomes conditional and creative, offering us valuable lessons. As Matsuda (1987: 324) argues, there is merit in looking to the social margins, 'adopting the perspective of those who have seen and felt the falsity of the liberal promise' to better understand the workings of law:

> [T]he best context within which to discuss ownership is not condominium development or ownership of valuable air space, but ownership as it appears in its absence, in confrontation with poverty, slavery, or unlawful occupation – property on the margins.
>
> (Van der Walt 2010: 90)

Scholarship and activism on commoning more generally can learn from the margins, informed and shaped by the brave defenders of the Downtown Eastside, 'rebuking the system/and speaking hope and possibility into situations/of apparent impossibility' (Osborn 1998: 282).

Notes

1 Versions of this chapter have been presented at a panel on C.B. Macpherson at Simon Fraser University, and a session at the Association of Law, Property and Society Conference in Vancouver, both in early 2014, before the September 2014 symposium at Cambridge from which this book derives. I am also grateful for the comments of Dave Diewert, one of the authors of the 'Zones of Exclusion' report noted here, as well as Frank Cunningham, Jeff Masuda, Sheila Foster, Alexandra Flynn, Brenna Bhandar, Noah Quastel, Liesl Spencer, Benjamin Niemark, and Lucy Finchett-Maddock.
2 Personal communication, Elvin Wyly, derived from the 2007 US Annual Housing Survey.
3 The Downtown Eastside currently contains 18,000 people, with some 5,000 living in private-sector single-room occupancy (SRO) housing. Many SRO rents have already increased above welfare rates. More than half of the population is low income, and nearly 90 per cent are renters. Ten per cent of the city's aboriginal population live in the neighbourhood.
4 It is useful to compare this to the community based 'right to remain' project within the Downtown Eastside, that similarly seeks to connect many of the historical layers of dispossession and struggle, see: www.revitalizingjapantown.ca/r2r/ (accessed 18 January 2016).
5 See: www.rennie.com/rms/news/woodwards-sets-sales-record-westbank-peterson-rennie/68#.VU6MBVy6yf8 (accessed 10 May 2015).
6 See: https://sites.google.com/site/zonesofex/ (accessed 21 October 2013).

7   See: http://ccapvancouver.wordpress.com/2013/01/28/sjzone/ (accessed 18 January 2016).

8   See: https://ccapvancouver.wordpress.com/2012/04/04/herb-varleys-speech-at-city-council-2/ (accessed 18 January 2016).

9   See: http://ccapvancouver.wordpress.com/2013/01/28/sjzone/ (accessed 18 January 2016).

10   Although it is heartening to see that Oxford University Press has recently reissued his major works, with new introductions by Frank Cunningham.

11   Moreover, there may be value in using Macpherson to complicate the prevalent collective/individual binary present in both critical and mainstream accounts of private and common property. Commons scholarship can fetishise 'community' in worrisome ways. There is also merit in socialising 'private' property.

12   This view also provides a useful counter to the influential view of property (including common property) as predicated solely, and appropriately, on exclusion (Merrill 1998).

13   Private property may also be thought of in this way, including the attempted erasure of common property (Blomley 2013).

14   We should also be careful in being overly nostalgic for the traditional commons. It was often highly exclusive, with the anxiety concerning 'masterless men' encroaching on local commons as comparable to contemporary middle-class anxieties concerning homeless people in public space (Hindle 1999).

15   Dagan similarly resists a singular view of the ends of property:

> Trying to impose a uniform conception of property on [the] diverse property institutions, which enable diverse forms of association and therefore diverse forms of goods to flourish, would be unfortunate because ... it would undermine the freedom-enhancing pluralism and the individuality-enhancing multiplicity so crucial to the liberal ideal of justice.
>
> (Dagan 2011: 43)

16   'Rather than a crushing limitation on liberty imposed by a powerful state, property law builds the floor on which we stand; it is the foundation that lets us live our lives in conditions of human decency' (Singer 2009: 1062).

References

Alexander, Gregory and Eduardo M. Peñalver. 'Properties of Community'. *Theoretical Inquiries in Law* 10, no. 1 (2009): 127–60.

Blackmar, Elizabeth. 'Appropriating "the Commons": The Tragedy of Property Rights discourse', in Setha Low and Neil Smith (eds), *The Politics of Public Space*. New York: Routledge, 2006, 49–80.

Blomley, Nicholas. 'Landscapes of Property'. *Law and Society Review* 32, no. 3 (1998): 567–612.

Blomley, Nicholas. *Unsettling the City: Urban Land and the Politics of Property*. New York: Routledge, 2004.

Blomley, Nicholas. 'Enclosure, Common Right and the Property of the Poor'. *Social and Legal Studies* 17 (2008): 311–31.

Blomley, Nicholas. 'Performing Property, Making the World'. *Canadian Journal of Law and Jurisprudence* 26, no. 1 (2013): 23–48.

Blomley, Nicholas. 'The Territory of Property'. *Progress in Human Geography*. Ahead-of-print publication 30 July 2015, doi:10.1177/0309132515596380.

Bollier, David and Silke Helfrich (eds). *The Wealth of the Commons: A World Beyond Market and State*. Amherst, NY: Levellers Press, 2012.

Bresnihan, Patrick and Michael Byrne. 'Escape into the City: Everyday Practices of Commoning and the Production of Urban Space in Dublin'. *Antipode* 47, no. 1 (2015): 36–54.

Carter, Adam. 'Of Property and the Human; or, C.B. Macpherson, Samuel Hearne, and Contemporary Theory'. *University of Toronto Quarterly* 74, no. 3 (2005): 829–44.

Dagan, Hanoch. *Property: Values and Institutions*. Oxford: Oxford University Press, 2011.

Garnett, Nicole S. 'Managing the Urban Commons'. *University of Pennsylvania Law Review* 160 (2012): 1995–2027.

Hartman, Chester. *Between Eminence and Notoriety: Four Decades of Radical Urban Planning*. New Brunswick, NJ: Rutgers Center for Urban Policy Research, 2002.

Hartman, Chester, Dennis Keating, and Richard LeGates. *Displacement: How to Fight It*. Berkeley, CA: National Housing Law Project, 1982.

Harvey, David. 'The Right to the City'. *New Left Review* 53 (2008): 23–40.

Harvey, David. *Rebel Cities*. London: Verso, 2012.

Helfrich, Silke. 'The Logic of the Commons and the Market: A Shorthand Comparison of their Core Beliefs', in David Bollier and Silke Helfrich (eds), *The Wealth of the Commons: A World beyond Market and State*. Amherst, NY: Levellers Press, 2012, 35–6.

Hindle, Stuart. 'Hierarchy and Community in the Elizabethan Parish: The Swallowfield Articles of 1596'. *The Historical Journal* 42, no. 3 (1999): 835–51.

Hodkinson, Stuart. 'The New Urban Enclosures'. *City* 16, no. 5 (2012): 500–18.

Huron, Amanda. 'Working with Strangers in Saturated Space: Reclaiming and Maintaining the Urban Commons'. *Antipode* (2015): doi: 10.1111/anti.12141.

Imbroscio, David L. 'Can we Grant a Right to Place?' *Politics & Society* 32, no. 4 (2004): 575–609.

Jeffrey, Alex, Colin McFarlane, and Alex Vasudevan. 'Rethinking Enclosure: Space, Subjectivity and the Commons'. *Antipode* 44, no. 4 (2012): 1247–67.

King, Peter. 'Gleaners, Farmers and the Failure of Legal Sanctions in England 1750–1850'. *Past and Present* 125, November (1989): 116–50.

Krueckeberg, Donald A. 'The Difficult Character of Property: To Whom Do Things Belong?' *Journal of the American Planning Association* 61, no. 3 (1995): 301–9.

Lang, Ursula. 'The Common Life of Yards'. *Urban Geography* 35, no. 6 (2014): 852–69.

Lee, Shin and Chris Webster. 'Enclosure of the Urban Commons'. *GeoJournal* 66, no. 1–2 (2006): 27–42.

Leiss, William. *C.B. Macpherson: Dilemmas of Liberalism and Socialism*. Montreal: New World Perspectives, 1988.

Lindsay, Peter. *Creative Individualism: The Democratic Vision of C.B. Macpherson*. Albany, NY: State University of New York Press, 1996.

Linebaugh, Peter. *The Magna Carta Manifesto: Liberties and Commons for All*. Berkeley, CA: University of California Press, 2008.

Macpherson, Crawford B. *The Political Theory of Possessive Individualism: Hobbes to Locke*. Oxford: Clarendon Press, 1962.

Macpherson, Crawford B. *Democratic Theory: Essays in Retrieval*. Oxford: Clarendon Press, 1973.

Macpherson, Crawford B. 'Humanist Democracy and Elusive Marxism: A Response to Minogue and Svacek'. *Canadian Journal of Political Science* 9, no. 3 (1976): 423–30.

Macpherson, Crawford B, (ed.). *Property: Mainstream and Critical Positions. Introductory and Concluding Essays by C.B. Macpherson*. Toronto: University of Toronto Press, 1978.

Macpherson, Crawford B. *Democratic Theory: Essays in Retrieval*. Oxford: Oxford University Press, 2012 [First published 1973].

Macpherson, Crawford B. *The Rise and Fall of Economic Justice*. Oxford: Oxford University Press, 2013 [First published 1984].

Maeckelbergh, Marianne. 'Mobilizing to Stay Put: Housing Struggles in New York City'. *International Journal of Urban and Regional Research* 36, no. 4 (2012): 655–73.

Matsuda, Mari. 'Looking to the Bottom: Critical Legal Studies and Reparations'. *Harvard Civil Rights – Civil Liberties Law Review* 22 (1987) 323–99.

Marris, Peter. *Loss and Change*. Revised edition. London: Routledge & Kegan Paul, 1986.

Merrill, Thomas W. 'Property and the Right to Exclude'. *Nebraska Law Review* 77 (1998): 730–55.

Nussbaum, Martha. 'Capabilities and Human Rights'. *Fordham Law Review* 66 (1997): 237–300.

Osborn, Bud. 'Raise Shit: Downtown Eastside Poem of Resistance'. *Environment and Planning D: Society and Space* 16, no. 3 (1998): 280–8.

Osborn, Bud. *Hundred Block Rock*. Vancouver: Arsenal Pulp Press, 1999.

Peñalver, Eduardo Moisés and Sonia K. Katyal. 'Property outlaws'. *University of Pennsylvania Law Review* 155 (2007): 1095–186.

Pierson, Christopher. *Just Property: A History in the Latin West. Volume One: Wealth, Virtue, and the Law*. Oxford: Oxford University Press, 2013.

Purdy, Jedediah. 'A Freedom-Promoting Approach to Property: A Renewed Tradition for New Debates'. *The University of Chicago Law Review* 72, no. 4 (2005): 1237–98.

Purser, Gretchen. 'The Circle of Dispossession: Evicting the Urban Poor in Baltimore'. *Critical Sociology*, ahead-of-print publication, 27 May 2014, doi: 10.1177/0896920514524606.

Quilligan, James B. 'Why Distinguish Common Goods from Public Goods', in David Bollier and Silke Helfrich (eds), *The Wealth of the Commons: A World beyond Market and State*. Amherst, NY: Levellers Press, 2012, 73–81.

Singer, Joseph W. 'Democratic Estates: Property Law in a Free and Democratic Society'. *Cornell Law Review* 94 (2009): 1009–62.

Tawney, Richard H. *The Agrarian Problem in the Sixteenth Century*. New York: Burt Franklin, 1961 [1912].

Strabowski, Filip. 'New Build Gentrification and the Everyday Displacement of Polish Immigrant Tenants in Greenpoint, Brooklyn'. *Antipode* 46, no. 3 (2014): 794–815.

Van der Walt, Andre. 'Property and Marginality', in Gregory S. Alexander and Eduardo M. Peñalver (eds), *Property and Community*. Oxford: Oxford University Press, 2010, 81–105.

Wenar, Leif. 'Rights'. *Stanford Encyclopedia of Philosophy*, edited by Edward N. Zalta, Fall 2011, at: http://plato.stanford.edu/archives/fall2011/entries/rights (accessed 18 January 2016).

White, James Boyd. *Justice as Translation: An Essay in Cultural and Legal Criticism*. Chicago, IL: Chicago University Press, 1990.

7 International humanitarian law and the possibility of the commons

Alex Jeffrey

Introduction

There are a number of immediate tensions involved in any attempt to entwine law and the commons. In many respects the terms work in opposition to each other, for where law operates as a tool of, first, imperialism and, second, capitalism, it is central to processes that have divided, converted, and undermined various forms of public life (Blomley 2008; Comaroff and Comaroff 2006; Jeffrey *et al.* 2012). But if work exploring the commons has conventionally centred on the relationship between humans and environmental resource allocation (see Bakker 2007), law has also been a fundamental framework through which claims to just access and allocation have been made. The central function of law, as Hyde (1997: 48) sets out, is to provide a 'totalising vocabulary' through which people can conceptualise their rights in a standardised fashion. While Hyde goes on to question and subvert this sense of totalisation, this discourse of universalism, where law is imagined as immutable over time (as a rule of law) and over space (as a jurisdiction), has proved central to the assertion of legal legitimacy over time (Valverde 2009). On the surface, this deviates some distance from a Marxist-inspired notion of the commons as the antithesis of capitalist enclosure, a situation wherein regimes of private property and individual obligation to the state crowd out other forms or spaces of common ownership and control (see Blomley 2014, 2015). But in this understanding of law there is also a minimalist interpretation of a legal commons centring on the experience of being bound to a common set of *rules*. Such rules nevertheless set to work on different spatial scales, so while law becomes the underpinning of certain *state*-conferred citizenship rights, it is also the means through which *global* commons (whether territorial or rights-based) may be protected.

This tension at the heart of law – between a repressive function of control and an emancipatory force of protecting rights and incursions – can be traced historically, as in the pioneering work of E.P. Thompson, which attests both to the savagery of law (particularly in its criminalisation of 'commoning') and to its radical potential (Thompson 1975). The establishment of common law in Europe in the twelfth and thirteenth centuries points to this duality. As responsibility for the adjudication of law shifted from local county courts to more centralised royal authority, so the more arbitrary elements of ecclesiastical jurisdiction were removed in favour of a standardised jury system. The perceived virtues of this system are clear: the delivery of justice becomes (more) consistent and the arbitration of guilt or innocence is the responsibility of fellow citizens. But as Blomley (2003) notes, the exercise of law always involves the exercise of violence (whether actual or implied, physical or symbolic) and the establishment of common law in England was a means through which new forms of monarchic power could be exercised in the nascent polity, a process that led on to the enclosure of common lands and the assertion of class-based hierarchies. Comaroff and Comaroff (2006) trace similar tensions between the supposed standardisation of law and the entrenchment of practices of dispossession within – and beyond – post-colonial states. There are, then, both interpretive and normative problems. The standardisation of law, the regulation of legal practice, and the participation of citizens in arbitration all gesture at a legal system that is displaying elements of a communal enterprise. The rule of law may thus point to a sense of a 'legal commons'. But at the same time, the exercise of law always unfolds within established power hierarchies and inevitably asserts the legitimacy of ruling elites (Jones 2015). In this sense, the creation of the legal system may be at the same time a repressive process that reproduces existing patterns of authority.

This dialectic of democratic and repressive functions has also been traced through the spatial attributes of law-making, trial justice, and recent work on 'lawfare' (Delaney 2015; Gregory 2010; Jones 2015). Any legal system requires a sense of the spatial limits of jurisdiction, thus suggesting a uniform plane within which certain bodies and objects are subject to a specific legal regime (Valverde 2009). This simple framework is, however, troubled by the more intricate geographies through which law achieves its authority. The enactment of law depends upon the division and bounding of space, the careful surveying of territories and bodies within a legal system and internal spatial divisions to demarcate different property or rights claims (Braverman *et al.* 2014). Such analysis prompts reflections on the specific – and asymmetric – geopolitical histories of law, in particular illuminating the colonial legacies that are masked by attempts to convey the neutrality and universalism of international law (Anghie 2007; Jones 2015). But alongside such spatial practices sits the enactment of law itself, a set of performances that have – since the classical regimes of Greek and Roman law – sought to use space as a means through which the legitimacy of legal deliberation may be conveyed. From the public agora through to Richard Rogers's European Court of Human Rights, legal practices have been imagined as public practices that adjudicate on behalf of – and in sight of – a particular collective (Mulcahy 2010). In both senses, law emerges less as an abstract articulation of a particular social order and more as a set of grounded and unfolding practices that are continually asserted and reworked.

This chapter seeks to continue this work examining the critical geographies of law through an exploration of recent legal innovations establishing institutions trying individuals for breaches of international humanitarian law (IHL). This body of law refers to instances where the norms of war, as laid out in the 1949

Geneva Conventions, are violated (O'Brien 1993). The geography of IHL has perhaps attracted less scholarly interest than other aspects of law, in part a reflection of the primacy of state-based legal systems and the lack of clear institutionalisation of international legal practices (Pearson 2008; Silbey 1997). As with other areas of international relations, the study of international law has often been caught in a 'territorial trap' whereby spatial intricacies are aggregated to competing state interests within the international system (Agnew 1994). But the purpose of this chapter is not simply to add spatial nuance to critical understandings of international law. Rather, I am seeking to focus on one particular element of the international legal system: whether we can see in the operation of IHL the basis for a revived notion of the *commons*. In doing so, the very conceptualisation of the commons comes into renewed focus. In spatial terms, the invocation of IHL is suggestive of a *planetary* sense of rights conferred by our common species membership, coupled with novel coalitions that operate transnationally to agitate for legal redress where violations of IHL have taken place. But the implications of law cannot be contained within the operation of international jurisprudence (see Boyle and Kobayashi 2015). The institutionalisation of IHL has also orientated attention to the role and significance of testimony, and in particular to the extent to which it is possible to bear witness to traumatic events and forge connections between violence in disparate parts of the planet. Finally, the operation of law, its materiality, performance and outcomes, has required new mobilisations, in terms of both expert interpretation of evidence and the enrolment of purportedly non-legal actors into juridical processes. These actions have blurred the distinction between law and non-law, where pressure is exerted through new transnational coalitions of knowledge production and innovative forms of civic action outside the legal arena. Following these disparate articulations of IHL and in line with the expanded conception of the commons set out in the introduction to this volume, I argue in this chapter that the possibility of revived legal commons can be analysed though three modalities: rights, testimony, and mobilisation. I argue that each modality works within and beyond law to evoke new commonalities that resist either a reduction to identity politics or the entrenchment of difference. Certainly, this formulation is at some remove from more materialist understandings of the commons, but the argument instead orientates attention to what is held *in common* through the operation of law, and the resources this provides for challenging violations of human rights.

The possibility of a legal commons, then, rests on an appreciation of the fragility and excesses of law – viewing law as a precarious achievement and one whose social effects always extend beyond the juridical. This argument builds on previous work, with Colin McFarlane and Alex Vasudevan (see Jeffrey *et al.* 2012), where the relationship between enclosure and the commons (or 'commoning') were understood to operate in a dialectic, arguing that a resurgent form of neoliberal geopolitics is producing possibilities for new forms of collective practice. If law is understood as a form of enclosure on more open-ended and socially-mediated forms of conflict resolution, so we simultaneously see potential for progressive interventions that widen access to rights and potentially foster new forms of justice. This argument is supported through qualitative field-work examining the establishment of war crimes trials at the Court of Bosnia and Herzegovina in Sarajevo undertaken between 2005 and 2014. This initiative marked the commencement of a transfer of legal responsibility for war crime from the International Criminal Tribunal for the former Yugoslavia (ICTY) in the Hague to the court in Sarajevo. International intervening agencies framed this manoeuvre as a democratisation of the war crimes process through its 'local-isation' to a state court (see Jeffrey and Jakala 2015). But in contrast to this rather straightforward legal geography, the establishment of the court produced varied effects, at once entrenching transnational sovereignty over the legal process while also cultivating collective actions calling for justice for crimes of the past. The court, then, sits at the heart of the enclosure-commons dialectic, illuminating the plural modalities of the commons produced through attempts to institutionalise elements of IHL.

The tensions of humanitarian ethics

It is difficult to pinpoint a precise starting date for the institutionalisation of IHL, though many commentators suggest either the Leipzig Trials in the wake of the First World War or the Nuremberg Trials following the Second World War (see, for example, Futamura 2007; Teitel 2003). The latter was particularly significant in provoking debate concerning transnational jurisprudence, though rather than framed in terms of jurisdiction (or the spatial limits of legal authority) these debates centred on the temporality of law. Legal positivists argued that adher-ence to the rule of law included recognition of antecedent law as valid, hence if the actions during the Second World War were legal under Nazi law, there are no grounds to retrospectively undertake criminal proceedings. In contrast, legal idealists promoted a concept of substantive justice, where certain crimes existed that were of such severity that they necessitated breaking with prior (in this case Nazi) legal systems (see Teitel 2000: 13). This friction between legal positivism (the absolute authority of existing rule of law) and idealism (the existence of 'crimes against humanity' that transcend any single jurisdiction) has shaped the subsequent form and discursive context of international courts and tribunals. In particular, political elites (at both global and national scales) have sought to carefully proscribe when and where rule of law is absolute, and conversely identify which military or paramilitary actions can be considered 'crimes against humanity': positivist legal formalism for some, idealist humanitarianism for others. The field of critical geopolitics is thus replete with studies examining the flexible usage of juridical instruments after war, belying the imagined uni-versalism of the internationally-brokered Geneva Conventions of the 1940s (Jeffrey 2011; Jones 2015; Morrissey 2011).

Perhaps one of the most intriguing examples of such brokerage can be seen in the slow institutionalisation of ad hoc tribunals in the 1990s addressing war crimes, first, through the International Criminal Tribunal for the former Yugoslavia (ICTY) (established in 1993 by UN Security Council Resolution 827) and, second, through the International Criminal Tribunal for Rwanda (ICTR) (established in 1994 through UN Security Council Resolution 955). The temporality of this process is complex and contested: some view these initiatives as a reflection of a unipolar post-Cold War world where a sense of stability and shared humanity may flourish (Wohlforth 1999), others see this as a reflection of those 'technological, infrastructural and communications advances that facilitated a conception of a common human condition' (Jeffrey 2009: 389). While it is tempting to read off these innovations as a virtuous humanitarian solidarity, internal reports of the conflicts within the United Nations surrounding the establishment of the ICTY and ICTR suggest otherwise. For example, Hazan (2004) charts the diplomatic manoeuvres by Western political elites to avoid general ethical and legal scrutiny of military operations, while also attempting to limit the implications of legal proceedings on the possibility of brokering peace. Such tensions are expressed through a political desire to show something being done (particularly in the case of the delayed and crisis-ridden intervention in the conflict in Bosnia and Herzegovina, 1992–95) while retaining a pragmatic desire to avoid indicting the interlocutors of peace agreements. From the outset, then, we see two tracks, one an imagined idealist position centring on a universal humanitarian ethics (something must be done when facing accusation of genocide) while also exhibiting Realpolitik: avoiding lengthy military engagements or hampering the resolution of the conflict.

The ethics of humanitarianism reflect this tension between idealism and pragmatism. In some respects, the establishment of ad hoc tribunals and juridical response to war crime seem to cohere with a Kantian ethics where the virtues of thinking beyond narrow state interests may provide the conceptual basis to limit the excesses of sovereign violence (Neiman 2008). But such species-level concerns should not be read off as a move to a more virtuously or morally-attuned foreign policy, rather it is a manoeuvre that must be set with the geopolitical frameworks that shape the colonial and post-Cold War world. It seems appropriate, therefore, that Didier Fassin (2012: 8) should draw a distinction between 'humanitarian morals (the principle on which actions are based and justified) and humanitarian politics (the implementation of these actions)'. This helpful dichotomy is one that appears in much writing on the validity of a Kantian humanitarian ethics: between the ideals and the practice, or what Seyla Benhabib (2004: 7) calls 'the growing normative incongruities between international human rights norms, particularly as they pertain to the "rights of others" ... and assertions of territorial sovereignty'. In a sense – and following the path set down by Fassin – the task at hand is to explore the relationship between these two fields: or, in other words, the ways in which the imaginaries of humanitarian ethics are institutionalised and practised in particular local settings and what political effects.

The potential distance between the ideals and practices of humanitarianism has provoked a range of scholarly enquires that seek to ground humanitarianism in actual existing political interventions. Chomsky's (1999) account of the 1999 NATO intervention in Kosovo stands as a critical touchstone for such work, a text that sought to challenge the accounts of London and Washington on the official purpose of the NATO intervention in Kosovo (a response to Serb atrocities in Kosovo), arguing instead that the intervention reflected a new unipolar world where the US-led NATO would project its power through military adventurism. Here, humanitarianism stands as a virtuous discursive frame, masking narrower sovereign interests pursued through aerial bombardment of an impoverished nascent state. Weizman's (2012) account of the ethics of modern military intervention makes a similar claim, though he traces this through a thickening web of agencies, norms, values and materials. For Weizman, humanitarianism, human rights, and international humanitarian law (IHL), 'when abused by state, supra-state and military action, have become the crucial means by which the economy of violence is calculated and managed' (2012: 3–4). The monitoring of human life, calculating its vulnerabilities, and weighing these into the nature and form of military incursions is the hallmark of what Weizman terms 'the humanitarian present'. The enactment of IHL lies at the heart of these processes, not least its emphasis on the proportionality of violence enacted during conflict. Weizman draws attention to two significant challenges in the implementation of IHL. First, that the upholding of IHL does nothing to try and limit war itself, it is directed solely at the appropriate forms of conduct during war. Second, the principle of proportionality provides 'no scale, no formulas and no numerical thresholds […] it demands assessment on a case-by-case basis, within parameters that are always relative, situational and immanent' (Weizman 2012: 12).

In the absence of immutable frameworks guiding the implementation of IHL there is a requirement to consider the moments and institutions through which such humanitarian law has been implemented and with what effects. Though a seeming contradiction, many of the studies tracing such implementations have focused their attention away from the process of law itself in order to look at the social contexts, political frameworks, and inter-subjective relationships through which legal processes operate (see Hughes 2015; Jeffrey and Jakala 2015; McEvoy 2007). Reflecting this approach, Boyle and Kobayashi (2015) explore the outcomes of people-led war crimes tribunals (PLWCTs) and in particular the Russell Tribunal 1966–67 set up to debate the actions of the US Government in Vietnam. This study does not centre on the legal implications of this initiative, but rather on the resources and ethical implications of attempting an extra-juridical justice instrument. This analysis deliberately oscillates at the meeting point of two interpretations of ethics, on the one hand a *humanitarian ethics* delivered by 'neutral, cool, and disembodied actors who apply universal particulars', and *care ethics* practised by 'engaged, emotional, and disembodied actors who reach judgements in the context of complex relationships' (Boyle and Kobayashi 2015: 14). But this distinction is difficult to sustain in practice and instead the forms of ethical judgement that emerge through the operation of PLWCTs is hybrid, a form of reasoning that rejects the colonial logics of the imagined-neutrality of Kantian ethics, but simultaneously resists a slide into ethical relativism. For Boyle and Kobayashi (2015) this necessitates a focus on the practical reasoning through which ethical judgement comes into being, a product of historical and geographical relationships traced on the ground and through situated human bodies.

Modalities of the legal commons

Tensions in the conceptualisation of humanitarian ethics point to the fact that if we are to trace a sense of the commons it will not be through an imagined unvariegated legal globalism but rather through the forms of social engagement and propinquity that are cultivated through the exercise of law (Amin 2004). Such an approach requires an exploration of the ways that juridical initiatives are reworked and contested by those subject to their jurisdiction, and – consequently – the ways in which care ethics, based on proximity, respect and civility flourish even in the shadow of juridical practice. Doing so brings to the fore the social relationships and patterns of authority that are reproduced and challenged through the practice of law, as Blomley makes clear, '[d]enaturalizing space [...] provides us with a way of resocializing law' (2003: 134). One of the first implications of 'denaturalizing' space is to challenge the concept of the legal process as removed from the social contexts within which it operates. While law gains much of its authority from a sense of abstract rationality, in practice it is a product of the entwining of legal and non-legal agents, materials, and institutions. In particular, the recent growth of forms of 'public outreach' from institutions such as the International Criminal Court, the ICTY, and the Court of Bosnia and Herzegovina (CBiH) to their constituents provides the opportunity to trace how law is spatialised and the forms of communal practices that are both envisaged and practised.

This widened perspective on the practice of law opens sightlines for rethinking the commons. Rather than a material resource, in this framework the commons stems from the more intangible set of 'hermeneutic resources' (Boyle and Kobayashi 2015: 13) produced through the operation of juridical processes. Not only does this point to a wider set of agencies beyond the narrow confines of legal action, it also encourages reflection on the spatial and temporal understandings of the outcomes of law. Rather than thinking of legal outcomes across the spatiality of particular jurisdictions and the temporality of trial justice, this approach attempts to think through the common resources produced through law and the implications for planetary understandings of justice and due process. But perhaps most significantly, the emergence of this legal commons is *hybrid* in nature, both incorporating legal rights and wider ethics of care and responsibility. In order to capture this hybridity, I have identified three potential modalities through which conceptions of the common, or commoning, may be traced: rights, testimony, and mobilisation.

Rights

The discourse surrounding IHL points firstly to a normative commitment towards legally-enshrined rights which are conferred on the basis of common humanity. This would seem to recall a sense of a natural law that stems from a common bodily form and cognition of an external world. But this imagined universalism also touches upon a more conceptually challenging notion: that of *cosmopolitanism.* For where common humanity points to the apparently 'natural' basis of shared legal system, the concept of cosmopolitanism points to the potential sense of shared civic and moral traits. Here we are not talking of cosmopolitan democracy, in the terms of Held (2009) or Archibugi (2003), but a return to the roots of cosmopolitanism as a means through which certain ideas of civility are enshrined as appropriate forms of public behaviour (Sennett 1976: 18). As Hyde notes,

> the experience of bodily self-control characteristic of modern notions of civility and politeness – these are not universal bodily experiences, but bodily practices with a distinct history – provide a [...] mental script for the abstract concepts of freedom and autonomy under law.
>
> (Hyde 1997: 51)

It is in such potentially emancipatory forms that we may connect the enforcement of IHL to the violence of neoliberal incursions, where global legal norms and frameworks provide the basis to resist the most crude forms of accumulation by dispossession, whether through dam construction, appropriation of commonly held resources, or privatisation of social housing (Blomley 2008: 324). While law is an important resource for such movements, it can produce a contradictory set of outcomes. For example, on numerous occasions the routing of the Israeli Wall around the West Bank has been subject to legal challenge at the Israeli High Court of Justice (HCJ) by Palestinian communities and Israeli civil rights groups. But as Weizman, drawing on Aeyal Gross (2006), has identified, such interventions serve establish a moral and judicial legitimacy for the projects as a whole. 'The "lesser evil" approach towards villagers,' continues Weizman (2007: 175), 'allowed a "greater evil" to be imposed on the Palestinian people as a whole.'

Similar uncertainties about the ethics of IHL have been expressed over the course of the research in BiH. For some, the court constituted an example of a Western-led intervention in the sovereign affairs of the Bosnian state, and consequently either withdrew involvement in legal processes (if a witness or victim) or took to national or social media to question the validity of the court's jurisdiction. While such actions often mimicked a critique of humanitarian ethics found in Kobayashi and Boyle's (2015) account of people-led war crimes tribunals, we must be careful to equate local resistance with a flourishing care ethics or even a positivist legal stance that questions the legitimacy of a new rule of law supplanting previous iterations. In a number of cases, resistance to the court was structured around the imagined legal competence of particular ethnic groups in the prosecution of IHL. For example, Milorad Dodik, the President of the Republika Srpska (RS), has successfully campaigned for a referendum (to be held in late 2015) for the RS to withdraw from the war crimes chamber within the CBiH. His reasoning stems from an imagined bias towards the crimes of Serbs and a consequent politicisation of IHL, arguing that:

the Court of BiH is under the direct influence of Bosnian politics. Our goal is to clearly specify the jurisdiction of the Court and the Prosecutor's Office [...] the Court and the Prosecutor's Office cannot have a monopoly on war crimes investigations and trials.

(Dodik 2015, author's translation)

While others viewed the court as a site through which some form of legal redress may be made for crimes of the past, the central concern of those in victims associations and human rights NGOs centred on either the ability of the court to complete the trial in a timely fashion or for the court to engage with the wider Bosnian public concerning its activities and legal outcomes. Many of the attempts to cultivate social engagement with international or humanitarian law are grounded in the language of *transparency*, either literally (through the use of glass walls to provide public surveillance of legal processes) or figuratively through the communication of legal processes to victims, witnesses, and wider community groups. The CBiH echoed such discourses of transparency, where the court presented its case outcomes through the publication of documents on a website:

So, then you can tell an eighty year old victim that everything is on a website [...] well, everything is on a website, what do you want? [Laughs] Well, I'm maybe being too harsh but that's what we sometimes [...] hear from the victim.

(Legal Advocacy NGO Representative, Sarajevo, 4 November 2011)

This interview excerpt captures the often disembodied nature of the outreach processes, where legal pronouncements were seen to have their effects through their iteration while the wider social context of their reception is considered either immaterial or beyond the concerns of legal performance. This attitude seems to capture what Sennett (1976: 27) refers to as the 'paradox of visibility and isolation', where the imagination of a visible public performance (albeit mediated through technology) is presented as a substitute for social interactions. In so doing, visibility and display become a proxy for participation. In these terms the concept of a common deliberation over law is structured as a trans-action between producers (performers) and recipients (audience) of legal practice.

This analysis also points to a decidedly emaciated form of common rights produced through the operation of the CBiH. From my own interview responses it was clear that the establishment of war crimes trials in BiH has created new opportunities for victims of the violence during the 1992–95 conflict to use jur-idical instruments to seek redress (see also Jeffrey and Jakala 2015; Nettelfield 2010). But while this process could be read in terms of a humanitarian ethics of universalising legal rights within a demarcated jurisdiction, the implementation of law has illuminated fissures and cracks in this apparently uniform plane. The entrenched ethnocracy (Toal and Dahlman 2011) that characterises many local governments in BiH allows an ethnic matrix to be super-imposed on the jurisdiction, thereby politicising the operation the court. At the same time, the court itself limits public engagement to the distribution of information, setting up a one-way stream of knowledge production, and limiting public interest in participation.

Testimony

The second potential modality of a legal commons examined in this chapter relates to the voicing of trauma and memory: the possibility of providing testimony. The crimes covered by IHL relate to violations of the customs of warfare, particularly relating to harm to civilian populations (for discussion, see Meron 1996). As discussed above, legal redress for contraventions of IHL have been heard through a range of institutions of transitional justice over the course of the last century, and while these institutions have advanced international jurisprudence, they have also created arenas through which testimony has been voiced and recorded. Such archives of memory constitute moments where sovereign power may be challenged by those subject to its violence (Edkins 2003), though the ability to voice testimony is constrained by both the embodied politics of memory and the mediating force of the juridical process, each pointing to a 'complex relationship between enunciation, listening and truth' (Hirsch and Spitzer 2009: 152). This complexity stems from the wider question that has stalked Holocaust and memory studies over the past fifty years: is it possible to bear witness – and provide testimony – under conditions where the terms to do not exist to articulate the trauma experienced (Levi 1989)? In more prosaic terms, the struggles to provide testimony also reflect the corporeal and material realities of law, where the structure of the court room, the provision of psychological support, or the questioning under cross-examination shape the ability of an individual witness to either provide testimony or to be viewed as a credible witness (Arendt 1994 [1963]; Felman 2002; Jeffrey and Jakala 2014). More than simply providing greater nuance within court exchanges, this work has argued that the *theatricality* of law is central to the communication of legal authority (Arendt 1994 [1963]; Felman 2002); whether through the architecture and materials of legal buildings (Latour 2010; Mulcahy 2010), the embodied nature of legal deliberation (Hyde 1997), or the scripting of court room exchanges (Jeffrey and Jakala 2014). Advocating more situated and embodied accounts, scholars have investigated the role of individual comportment, assertions of masculinity and the body as a site of violence (and consequently a repository of evidence) within the practice of law (Clarkson 2014; Felman 2002; Hyde 1997). But such questions of intelligibility or credibility keep testimony within the narrow channel of the juridical, rather than viewing such acts as human enactments against barbarity, what has been termed by Hirsch and Spitzer as forging a 'cosmopolitan memory [...]' achieved when 'the uniqueness and exceptionalism attributed to [...] victim suffering for nationalist ends and the field of memory is broadened to include other victims, other perpetrators and other bystanders involved in acts of mass violence and persecution' (2009: 165). In this sense, testimony may be re-scaled from a juridical category of providing evidence, towards a humanitarian act challenging sovereign and colonial violence.

In the case of the CBiH, the need to provide testimony had both juridical and social consequence. In the trial process, the embodied nature of the provision of testimony came to the fore. The use of materials and bodily practices to challenge the production of testimony was highly gendered. The power asymmetry of the interrogations was acknowledged by the prosecutors, who felt concerned about the production of legal masculinities that thrived on the vulnerability of those testifying. During a public dialogue concerning trial practices in the northern Bosnian town of Tuzla, one prosecutor emphasised the incapacity of judges or prosecution teams to intervene and the potential for the cross-examination process to be wielded as a tool by defendants to unsettle the witness 'we are talking about very vulnerable individuals. During the interrogations verbal offenses are being used in order to disturb an individual and neither we nor the judge nor the prosecutor can help' (interview with a public prosecutor, Tuzla, 19 April 2012). The lack of protection for witnesses was an enduring theme of the interviews, in particular the absence of state-financed psychological assistance. Indeed, there was a lingering concern that the emotional fragility of the witness played a role in establishing the legal veracity of the testimony: 'I have a statement,' remarked a representative of a legal advocacy NGO in Sarajevo, 'by one of the women who said that they didn't believe her [about being raped] until she started crying' (Jeffrey and Jakala 2014: 664).

These accounts reinforce Felman's (2002) point that giving testimony may have *retraumatising* effects. This was one among a number of concerns that victims' associations voiced regarding giving testimony, alongside the fear of exposure within the community, and the concern that the process would be futile if the defendant was found either not guilty or given what would be perceived as a lenient sentence. But one of the key concerns aired related the possibility of shared position as a victim, of generalising trauma beyond an individual status. It seems that such an effect would be crucial to the constitution of Hirsch and Spitzer's (2006) concept of a 'cosmopolitan memory'. Instead, the narratives from the fieldwork spoke of the fragmentation and isolation of individuals, beyond their utility as a function within the juridical process. The President of the Association of Camp Detainees from the southern Bosnian town of Prozor commented on this isolation, 'how can we reach other individuals who have survived rape in Prozor, Mostar, Čapljina, Stolac and other places where it took place? [...] I think victims have no protection at all' (public dialogue submission, Mostar, 18 May 2012). By emphasising the legal – as opposed to social – practice of testimony, the forms of truth telling remained resolutely connected to their ability to account for the guilt or innocence of individuals within the war crimes trials. But this interpretation misses the role of testimony in consolidating a traumatised and divided polity. As Hartman (1995) suggests, the strength of testimony is not in providing 'truths' for a positivist history, but rather in 'recording the psychological and emotional milieu of the struggle for survival, not only then, but also now' (Hartman 1995 in Hirsch and Spitzer 2009: 155).

Mobilisation

The final modality of the legal commons explored in the chapter explores *mobilisations* and focuses attention on the excesses of law: namely, that the political and social effects of the juridical process always exceed the verdict and sentencing outcomes. This excess is a function of the materiality of law, where the possibility of law requires forms of mediation and knowledge creation that exist largely outside the control of legal instruments. For example, in a collaborative intervention Eyal Weizman (2014) has drawn attention to forensics as a deliberative field of knowledge production, where technologies, claims to expertise and historic knowledge hierarchies shape claims to legal truth. Consequently for Weizman and his collaborators studying forensic practices affords insights into the processes through which meaning and value are ascribed to objects within public deliberations. This is partly, then, a body of work that illuminates the complex material, bodily and affectual processes through which knowledge is stabilised within contested political or legal situations. In so doing this work extends recent interest in the materiality of knowledge controversies, where the properties, temporalities and mutability of materials shapes the nature of economic and political deliberation (see Barry 2013). But alongside this question of the constitution of expertise within knowledge controversies, work on forensics also highlights the contradictory nature of truth claims, where – for example – the precision of screen resolution, the temporality of pH tests, or the definition of appropriate force all shape the interpretation of material evidence. Highlighting this corporeal and material assemblage, its choreography in court space, and its implications for completing trial processes, illustrates how the completion of legal procedures requires the existence and compliance of a range of factors beyond the immediate control of court or legal officials. Whether the provision of testimony, the legibility of an identity card, or the location of mortal remains: law depends upon the enrolment of numerous 'message bearers' in the production of meaning (Whatmore 2002: 3). In so doing a straightforward notion of legal subjectivity is challenged in two respects. First, the role of decay and decomposition illuminates forms of non-human agency, processes that Caitlin DeSilvey (2006: 323) notes may not be summarily dismissed as 'erasure' but rather as 'generative of a different kind of knowledge'. One of the tangible outcomes of decay is, for instance, the incompletion of trial processes as evidence is incomplete or illegible. Second, this focus on materiality has foregrounded the role of mediating agents in narrating the significance of objects and materials. As Weizman (2014) makes clear, this opens the space for transnational movements that challenge official centres of knowledge production and provide counter-forensics that illuminate abuses of power and violence against humanity.

This sense of plural agency and excess was reflected in trials monitored as part of the research in BiH. The circulation of materials within the courtroom shaped claims to truth, most explicitly through the presentation of material evidence. But rather than deliberating forensic science as captured in Weizman's

(2014) account, the discussion of evidence often related to prosaic questions of whether or not the documents contained an administrative stamp or the date on which a document was certified. On one occasion, for example, the Prosecutor's witness was presented by the defence lawyer with a map of BiH and asked to state the acreage of the Srebrenica enclave and then outline it on the map. The map used was in fact a poor quality computer printout reproduced on an A4 paper with no scale to help determine the acreage. The witness stated that the only thing he could ascertain from the map was that it was of BiH and he would be unable to determine the acreage of the enclave as there was no scale. As the witness clutched the crumpled map in confusion, the defence did not offer an alternative document, indeed there was a large map of BiH on the wall of the courtroom that could have been used (court observations, 1 November 2011). Whether the map was deliberately of poor quality in order to cast doubts on the claims of the witness, or whether conversely the map's quality was an indication of the bureaucratic inadequacies of the defence team is open to debate. What is certain is that the map itself was enrolled into the bodily contestation between defence and witness; a material extension of the anxieties and struggles for claims to truth.

While the map example highlighted the significance of material evidence, the later sessions in the same trial pointed to the forms of DeSilvey's (2006) non-human agency that may be at work when considering the evidential processes within war crimes trials. In the following session of the same trial the Defence Team offered a series of further submissions of material evidence, largely comprising a variety of identification cards that sought to illustrate their client's status as a prisoner of war. The document that elicited the most scrutiny was the fake military identification issued during the war in Bijeljina, BiH. A significant amount of time was spent examining the Bijeljina stamp dated 8 December 1994 to determine whether or not it was a real stamp from the war government. The fragment of paper had a particular significance to the room, it was carefully considered by defence and prosecution lawyers, as each tried to gauge whether the stamp was genuine, carefully studying the faded and worn document. The deterioration was a product of the passage of time, it had been stored in the defendant's house and it had become bleached by the sun and crumpled against other papers. At the conclusion of the discussion the card was carefully placed on a scanner to be recorded electronically – frozen in time – and conveniently accessible for potential appeals. The incident underscored the point that each piece of evidence is itself decaying, slowly decomposing as it is exposed to atmosphere, buried in the soil or hidden in among mundane objects in potential defendants' homes.

The fragility of evidence has only increased the need for human testimony to complete trials. But, as we have seen, the provision of testimony is fraught and often unpopular with those called upon to bear witness. In these circumstances the court has turned to human rights NGOs and other civil society organisations to form partnerships that may assist in the provision of evidence and testimony (see Jeffrey 2011). These practices of 'public outreach' involve seminars concerning trial outcomes, workshops with victims' associations, and education programmes in schools and university. It is clear from the research that there are profound social implications of public outreach, since each of the organisations or individuals involved actively sought to table alternative or additional (often restorative or therapeutic) elements into legal processes. There were moments of exchange, crisis, and breakdown as sentencing workshops (where the court justified the sentences given to convicted war criminals) or outreach seminars (where victim associations and court officials met) provided sites of interaction, often spilling into local cafés and bars afterwards. Representing a marked difference to the rather dry exchanges of court space, these were moments of high emotion, where those who had either participated in a trial process, or were keen to see a particular individual face justice met with court officials, political elites, and members of international organisations. Examining these elements of practice requires studying the actions not of courts but rather the more informal and often overlooked practices of the wider set of civil society agencies. This approach captures the fact that the organisations and individuals involved were calculating and effective; they were not mere dupes in a rigid legal performance. Rather, the outreach process opened up coalitions and opportunities that were seized upon by organisations to create spaces for public dialogue concerning justice.

This interplay between legal structure and associative agency could be explained as a form of an embodied legal habitus, where primacy of law is internalised by individual agents, while incorporating a sense of the fragility of this system and the possibilities for alternatives. Such resistance in these terms is not about attempting to halt the legal process, but rather attempts to illuminate and practice spheres of activity and deliberation that are absent from law. For example, the research examined the work of a victims' association in north-west Bosnia and Herzegovina which was engaged in public outreach but was also aware of the limitations of this process in fostering public debate. As a corrective, the organisation was undertaking a series of educational programmes designed to preserve public memory of the genocide that took place in their town. Books had been produced, an annual commemorative day established, and the public wearing of white armbands in the streets of the town as an indication of solidarity in the commemoration of violence of the past. The significant part of this latter protest was its claiming of space, the wearing of armbands was started as an act of public defiance at the reopening of the former camp at Omarska as a functioning corporate enterprise by Arce-lorMittal Steel Corporation. These attempts to solicit public engagement in questions of justice were not in direct conflict with the legal process, but they sought to highlight the potential silences within a retributive legal system to questions of commemoration, solidarity, and redress. Such actions are structured around a sense of agonistic engagement with a differentiated public, where different views of justice and law circulate within public space. Within such protests forms of dress, behaviour, and comportment cultivate debate as to the boundaries and constitution of community and jurisdiction in a divided post-conflict polity. While this is not a natural or inevitable outcome of international legal programmes, the actions of this group have been supported by the existence of legal arbitration over crimes committed in the past.

Conclusion

This chapter has considered how the practice of IHL can be interpreted as providing the material and immaterial resources to foster new practices of the commons. Prompted by work celebrating the emancipatory potential of forms of jurisdiction that exceed the territoriality of the state, this chapter has sought to refocus attention on the *practices* through which IHL (and by extension that of the commons) is realised. This argument can be inserted into a wider intellectual tradition that has examined the uneven exercise of seemingly totalising legal frameworks. In particular, the discussion has sketched the outlines of a tension between an imagination of IHL as a universal set of legal potentialities, and the seemingly partial and exclusive processes through which law unfolds. This finding underscores the point that access to legal redress is a function of social, cultural, and economic status. This sense of uneven access to justice is lost in an imaginary of legal communal practices structured around a sense of transparency, where the possibility to *see* legal processes serves merely as a substitute to any truly participatory forms of engagement. In making this point we must be careful not to construct a simple geometry between a virtue of participation and the dangers of transparency. As Oomen (2005) notes in the case of the *Gacaca* process in Rwanda, even where legal processes involve many community members, this can continue to reassert established social hierarchies and can result in individuals participating merely to fulfil obligations.

But this chapter also glimpses alternative modalities of the common that are provoked by, and exist alongside, these orthodox legal performances. Reflecting the main arguments made in the introduction to this book, such sightlines do not equate to a utopian and tangible commons, but are rather pointing to the forms of deliberative sites and discursive properties that are produced through the exercise of IHL. Unquestionably, the existence of war crimes trials in Sarajevo has provided the opportunity for individuals to seek legal redress for crimes committed against themselves or their families during the 1992–95 conflict, opportunities that were not open to them when the transitional justice process was the primary preserve of the ICTY. In addition, the demands to provide testimony are creating archives of material that may – at some future point – provide the 'hermeneutic resources' (Boyle and Kobayashi 2015) to challenge violent and discriminatory politics in the future. Perhaps most tangibly, the implementation of IHL has required the participation of a range of associations and non-governmental actors, both to complete the instrumental aspects of trial justice but also to communicate the nature and implications of law through wider social networks. These processes do not produce a tangible materiality we can classify uncritically as 'common', but they nevertheless seek to develop more public deliberations concerning inevitably divergent understandings of justice. In the contexts under examination in this chapter, this has involved challenging narrow, retributive forms of law not necessarily with a view to positing a straightforward alternative, but rather to encourage debate as to the nature of justice, memory, and community *in common*.

The centrality of public interactions to this notion of the commons returns the discussion to the question of *cosmopolitanism*. Such public debate requires social civility, where deliberation can take place without violence and with respect to difference. Where retributive forms of justice encourage testimony and the production of (often limited) archives of memory, they also provoke adversarial identity positions (between individualised victim and aggressor, for example). This conflict between the possibilities of debate and the rigidity of legal classification highlights the inherent symbolic violence of law, where subject positions are ascribed and fixed through the machinations of legal deliberation. In these terms it is the common experience of being subject to law, rather than differential subject positions within a legal process, that must be emphasised within any sense of a common legal enterprise. Hence the use of legal mechanisms to arbitrate on humanitarian crimes provides the sightlines towards more cosmopolitan forms of law, but such cosmopolitanism will always be 'to come' since these legal deliberations take place in the face of considerable violence and social fragmentation.

References

Agnew, John. 'The Territorial Trap: The Geographical Assumptions of International Relation Theory'. *Review of International Political Economy* 1, no. 1 (1994): 53–80.

Amin, Ash. 'Regions Unbound: Towards a New Politics of Place'. *Geografiska Annaler B* 86, no. 1, (2004): 33–44.

Anghie, Anthony. *Imperialism, Sovereignty and the Making of International Law*. Cambridge: Cambridge University Press, 2007.

Archibugi, Daniele. 'Cosmopolitical Democracy', in Daniele Archibugi (ed.), *Debating Cosmopolitics*. London: Verso, 2003, 1–15.

Arendt, Hannah. *Eichmann in Jerusalem: A Report on the Banality of Evil*. London: Penguin Books, 1994 [1963].

Bakker, Karen. 'The "Commons" Versus the "Commodity": Alter–Globalization, Anti-Privatization and the Human Right to Water in the Global South'. *Antipode* 39, no. 3 (2007): 430–55.

Barry, Andrew. *Material Politics: Disputes Along the Pipeline*. Chichester: Wiley-Blackwell, 2013.

Benhabib, Seyla. *The Rights of Others: Aliens, Residents, and Citizens*. Cambridge: Cambridge University Press, 2004.

Blomley, Nicholas. 'Law, Property, and the Geography of Violence: The Frontier, the Survey, and the Grid'. *Annals of the Association of American Geographers* 93, no. 1 (2003): 121–41.

Blomley, Nicholas. 'Making Space for Law', in Kevin Cox, Murray Low, and Jenny Robinson (eds), *The Sage Handbook of Political Geography*. London: Sage, 2008, 155–68.

Blomley, Nicholas. 'Making Space for Property'. *Annals of the Association of American Geographers* 104, no. 6 (2014): 1291–306.

Blomley, Nicholas. 'The Ties that Blind: Making Fee Simple in the British Columbia Treaty Process'. *Transactions of the Institute of British Geographers* 40, no. 2 (2015): 168–79.

Boyle, Mark and Audrey Kobayashi. 'In the Face of Epistemic Injustices? On the Meaning of People-Led War Crimes Tribunals'. *Environment and Planning D: Society and Space*, ahead-of-print publication 10 August 2015, doi: 10.1177/0263775815598101.

Braverman, Irus, Nicholas Blomley, David Delaney, and Alexandre Kedar (eds). *The Expanding Spaces of Law: A Timely Legal Geography*. Palo Alto, CA: Stanford University Press, 2014.

Chomsky, Noam. *The New Military Humanism: Lessons from Kosovo*. London: Pluto Press, 1999.

Clarkson, Carrol. *Drawing the Line: Toward an Aesthetics of Transitional Justice*. New York: Fordham University Press, 2014.

Comaroff, Jean and John Comaroff. 'Law and Disorder in the Postcolony: An Introduction', in Jean Comaroff and John Comaroff (eds), *Law and Disorder in the Postcolony*. Chicago, IL: University of Chicago Press, 2006, 1–56.

Delaney, David. 'Legal Geography I: Constitutivities, Complexities, and Contingencies'. *Progress in Human Geography* 39, no. 1 (2015): 96–102.

DeSilvey, Caitlin. 'Observing Decay: Telling Stories with Mutable Things'. *Journal of Material Culture* 1, no. 3 (2006): 318–38.

Dodik, Milorad. 'Dodik: Sud i Tužilaštvo BiH pod direktnim uticajem bosnjačke politike', *Nezavisne Novine*, 20 August 2015, at: www.nezavisne.com/novosti/bih/Dodik-Sud-i-Tuzilastvo-BiH-pod-direktnim-uticajem-bosnjacke-politike/321465 (accessed 27 August 2015).

Edkins, Jenny. *Trauma and the Memory of Politics*. Cambridge: Cambridge University Press, 2003.

Fassin, Didier. *Humanitarian Reason: A Moral History of the Present*. Translated by Rachel Gomme. Berkeley, CA: University of California Press, 2012.

Felman, Shoshana. *The Juridical Unconscious: Trial and Traumas in the Twentieth Century*. Cambridge, MA: Harvard University Press, 2002.

Futamura, Madoka. *War Crimes Tribunals and Transitional Justice: The Tokyo Trial and the Nuremburg Legacy*. Abingdon: Routledge, 2007.

Gregory, Derek. 'War and Peace'. *Transactions of the Institute of British Geographers* 35, no. 2 (2010): 154–86.

Gross, Aeyal. 'The Construction of a Wall between The Hague and Jerusalem: The Enforcement and Limits of Humanitarian Law and the Structure of Occupation'. *Leiden Journal of International Law* 19 (2006): 393–440.

Hartman, Geoffrey. 'Learning from Survivors: The Yale Testimony Project'. *Holocaust and Genocide Studies* 9, no. 2 (1995): 192–207.

Hazan, Pierre. *Justice in a Time of War: The True Story Behind the International Criminal Tribunal for the Former Yugoslavia*. College Station, AZ: Texas A&M University Press, 2004.

Held, David. 'Restructuring Global Governance: Cosmopolitanism, Democracy and the Global Order'. *Millenium: Journal of International Studies* 37 (2009): 535–47.

Hirsch, Marianne and Leo Spitzer. 'The Witness in the Archive: Holocaust Studies/Memory Studies'. *Memory Studies* 2, no. 2 (2009): 151–70.

Hughes, Rachel. 'Ordinary Theatre and Extraordinary Law at the Khmer Rouge Tribunal'. *Environment and Planning D: Society and Space*, ahead-of-print publication 10 August 2015, doi: 0263775815598081.

Hyde, Alan. *Bodies of Law*. Princeton, NJ: Princeton University Press, 1997.

Jeffrey, Alex. 'Justice Incomplete: Radovan Karadzic, the ICTY, and the Spaces of International Law'. *Environment and Planning D: Society and Space* 27, no. 3 (2009): 387–402.

Jeffrey, Alex. 'The Political Geographies of Transitional Justice'. *Transactions of the Institute of British Geographers* 36, no. 3 (2011): 344–59.

Jeffrey, Alex and Michaelina Jakala. 'The Hybrid Legal Geographies of a War Crimes Court'. *Annals of the Association of American Geographers* 104, no. 3 (2014): 652–67.

Jeffrey, Alex and Michaelina Jakala. 'Using Courts to Build States: The Competing Spaces of Citizenship in Transitional Justice Programmes'. *Political Geography* 47 (2015): 43–52.

Jeffrey, Alex, Colin McFarlane, and Alex Vasudevan. 'Rethinking Enclosure: Space, Subjectivity and the Commons'. *Antipode* 44, no. 4 (2012): 1247–67.

Jones, Craig. 'Lawfare and the Juridification of Late Modern War'. *Progress in Human Geography*, ahead-of-print publication 16 March 2015, doi: 10.1177/0309132515572270.

Latour, Bruno. *The Making of Law: An Ethnography of the Conseil d'Etat*. Cambridge: Polity Press, 2010.

Levi, Primo. *The Drowned and the Saved*. Translated by Raymond Rosenthal. New York: Vintage, 1989.

McEvoy, Kieran. 'Beyond Legalism: Towards a Thicker Understanding of Transitional Justice'. *Journal of Law and Society* 34, no. 4 (2007): 411–40.

Meron, Theodor. 'The Continuing Role of Custom in the Formation of International Humanitarian Law'. *American Journal of International Law* 90, no. 2 (1996): 238–49.

Morrissey, John. 'Liberal Lawfare and Biopolitics: US Juridical Warfare in the War on Terror'. *Geopolitics* 16, no. 2 (2011): 280–305.

Mulcahy, Linda. *Legal Architecture: Justice, Due Process and the Place of Law*. Abingdon: Routledge, 2010.

Neiman, Susan. *Moral Clarity*. Princeton, NJ: Princeton University Press, 2008.

Nettelfield, Lara. *Courting Democracy in Bosnia and Herzegovina: The Hague Tribunal's Impact in a Postwar State*. Cambridge: Cambridge University Press, 2010.

O'Brien, James. 'The International Tribunal for Violations of International Humanitarian Law in the Former Yugoslavia'. *American Journal of International Law* 87, no. 4 (1993): 639–59.

Oomen, Barbara. 'Donor-Driven Justice and its Discontents: The Case of Rwanda'. *Development and Change* 36, no. 5 (2005): 887–910.

Pearson, Zoe. 'Spaces of International Law'. *Griffith Law Review* 17, no. 2 (2008): 489–514.

Sennett, Richard. *The Fall of Public Man*. London: W.W. Norton & Company, 1976.

Silbey, Susan. '1996 Presidential Address. "Let Them Eat Cake": Globalization, Postmodern Colonialism, and the Possibilities of Justice'. *Law and Society Review* 31, no. 2 (1997): 207–35.

Teitel, Rudi. *Transitional Justice*. New York: Oxford University Press, 2000.

Teitel, Rudi. 'Transitional Justice Genealogy'. *Harvard Human Rights Journal* 69 (2003): 69–94.

Thompson, Edward Palmer. *Whigs and Hunters: The Origins of the Black Act*. New York: Pantheon, 1975.

Toal, Gerard and Carl Thor Dahlman. *Bosnia Remade: Ethnic Cleansing and its Reversal*. New York: Oxford University Press, 2011.

Valverde, Mariana. 'Jurisdiction and Scale: Legal Technicalities as Resources for Theory'. *Social & Legal Studies* 18, no. 2 (2009): 139–57.

Weizman, Eyal. *Hollow Land*. London: Verso Books, 2007.
Weizman, Eyal. *The Least of All Possible Evils: Humanitarian Violence from Arendt to Gaza*. London: Verso Books, 2012.
Weizman, Eyal. 'Introduction: Forensis', in Forensic Architecture (ed.), *Forensis: The Architecture of Public Truth*. Berlin: Sternberg Press and Forensic Architecture, 2014, 9–34.
Whatmore, Sarah. *Hybrid Geographies: Natures Cultures Spaces*. London: Sage, 2002.
Wohlforth, William. 'The Stability of a Unipolar World'. *International Security* 24, no. 1 (1999): 5–41.

8   The shrinking commons and uneven geographies of development[1]

Sarah A. Radcliffe

Introduction

Speaking at a 2014 meeting sponsored by the Inclusive Capitalism Initiative at the Mansion House in London, the Managing Director of the International Monetary Fund (IMF), Christine Lagarde, spoke about 'rising income inequality, and the dark shadow it casts across the global economy'. She noted how 'the richest 85 people in the world, who could fit into a single London double-decker [bus], control as much wealth as the poorest half of the global population, that is 3.5 billion people' (Lagarde 2014). Such comments raise questions about the nature of the proposed inclusive capitalism, and how it might shape hegemonic global relations with the world's poor – the 'base of the pyramid'. Meanwhile, thousands of miles away from the City of London, a German non-government organisation (NGO) was giving out small amounts of no-strings-attached cash to nearly 700 of Niger's households uprooted by food production collapse, food price rises, and conflict-displaced people. Handing out money (unconditional cash transfers in policy language) aimed to permit people to return to planting fields, alleviating longer-term hardship among households including those led by disabled women (CaLP: n.d.). Does this third sector initiative reflect inclusive capitalism, and how successfully does this model address the bottom billion, and on what terms?

Development's discursive and material interventions are premised, this chapter argues, on the fiction of a common goal of improvement in which richer countries and their delegated institutions have a responsibility to assist distant strangers and lessen the harshest impacts of skewed global wealth distributions.[2] This fiction is sustained by a vision of a common humanity such that when neediness arises, it elicits humanitarian acts and inter-subjective peer-to-peer assistance. However notwithstanding the sterling efforts of agencies, professionals and donations, billions of the world's population – largely, but not exclusively in the global South – experience the daily indignities of impoverishment, dispossession, and marginalisation (Fanon 2001 [1961]). Growing inequality coexists with (few) unconditional and (many, highly) conditional cash handouts, creating a commons-impoverishment paradox, a conjuncture of opposing processes by no means unique to the twenty-first century. As scholarship amply documents, development interventions play on the appeal of helping the neediest while bracketing off real consideration of the broader political economic and political processes (Ferguson 1990; Li 2007). Critical studies of humanitarianism demonstrate its close association with imperialism and its entanglement with Western religiosity and economics. Currently the IMF – and its Bretton Woods counterparts – continues to hold neoliberal sway, merely tinkering with social relations and targeting aid while leaving political debates around structural change out of the frame. Moreover, the affective content of an impulse to global generosity is largely premised on attributing vulnerability to the poorest, a sentimental reading that encodes and brackets off political questions. The chapter examines the configuration of today's commons-impoverishment paradox, its institutionalisation, and affective penumbra, in order to discern the parameters of (un-)commoning in the field of development (Amin and Howell: Chapter 1). I begin with an overview of developmental responses to 'common humanity' as well as global patterns of disengagement (Section I).[3] One key model for making capitalism inclusive is the practice of granting cash under strict *conditions*, involving ever-finer distinctions between subgroups of impoverished humanity, often on the basis of supposed vulnerability and non-capitalist positionalities. Moreover, such programmes are being funded just as public services and infrastructures are dismantled or delegitimised. The resulting paradox of a (fictional) commons combined with (material) impoverishment is thus in part realized through social protection, a global policy turn that seeks to bridge the incompatibility of pro-poor measures with a vigorous market economy (Section II). What I term the commons-impoverishment paradox refers then to how interventions lie at a point of tension between acknowledging growing inequality while speaking to world humanity, a shift from Cold War rhetoric. Ethnographic explorations of Latin American social protection offer specific insights into such dynamics. And yet, the fiction of an inclusive capitalism has, in recent years, been seriously challenged in Latin America's 'post-neoliberal' regimes and their efforts to reconfigure distribution, citizenship, poverty-alleviation, and an unjust political economy. Section III thus addresses the parameters and contradictions found in these innovative efforts to rethink a citizenship founded on a commons of rights, profit sharing, and universal social programmes.

I Disengagement

> Chronic poverty has been allowed to remain over time both by certain social norms and by becoming politically institutionalized in ways that ensure [its] reproduction ... over time.
>
> (Hickey and Bracking 2005: 855)

> If development can be seen as a formula for sharing the world with others, in its present configuration many seem destined to die before their time, while others are able to live beyond their means.
>
> (Duffield 2010: 57)

Notwithstanding global improvements in poverty-alleviation, child survival rates, enrolment in primary education, the access to water, and national debt reduction, the development commons remains cross-cut by inequality and unevenness. Large parts of the global South are becoming landscapes of abandonment for many and accumulation for a few. As traced in this section, the global 'development commons' are characterised by aid budgets that struggle to keep up with Foreign Direct Investment (FDI) and remittances, development as a form of securitisation, the creation of surplus populations and abandonment (characterised by racialisation, abjection, insecurity, and experimentation), the production of impairment and generationally-specific vulnerabilities, and affective disengagement.

The patterning of overseas development assistance (ODA) reveals the fragility of the economic basis for a development commons. ODA has been less impacted by the global economic crisis than might be expected, although its geopolitics and spatial distribution follow trends that contribute to growing inequalities in income and development. Foreign Direct Investment (FDI) has certainly outstripped the flows of overseas development aid through the 1990s and into the twenty-first century, while remittances were not far behind.[4] Moreover, aid is increasingly routed via the Bretton Woods Institutions, which accounted for one-third of total aid by the turn of the century despite its record for aid volatility. In combination, these processes lead to a 'regressive shift in the allocation of aid away from some of the poorest countries' (White and Feeny 2003: 133). The recent global crisis led to a nearly 3 per cent fall in aid in 2011, while the least developed countries saw a decline in bilateral ODA of nearly 9 per cent, particularly affecting Central America, the Philippines, and Indonesia.[5] OECD countries have however donated increased aid levels in 2013, although the trend in falling aid to sub-Saharan Africa looks likely to continue; for all the TV screens showing starving African children, ODA to sub-Saharan Africa declined through the 1990s, matched almost exactly by increased aid to Europe (White and Feeny 2003: 122).[6]

Likewise, donor nations' deliberate exit from middle-income countries (e.g. the UK was, for example, to end aid to India in 2015) contributes to notably dynamic geographies of aid and development. Overall, ODA has become increasingly conditional, selective (flowing to countries already showing their commitment and ability to fulfil donors' criteria), concentrated (individual donors give to fewer countries[7]), and volatile (subject to year-on-year fluctuations) (Hermes and Lensink 2001; White and Feeny 2003). Together with shifts in aid geographies, ODA remains intrinsically performative; 'interventionism appears cyclical, ongoing and expansive' as, for example, in the wake of the South Asian tsunami, shoring up a fragile impulse towards a common humanity (Duffield 2010: 56; Duffield and Hewitt 2009). In this sense, the paper-thin fiction of a development commons is reiterated every time a government misses its target GDP spending on aid, or reneges on a stated commitment to provide humanitarian relief.

In part this articulation of material retraction and spectacular commoning is grounded in a geopolitical Realpolitik whereby rich countries decide the terms upon which poorer countries and populations partake in global authority and resources. In its Cold War coding, development interventions built geopolitical allies, contained alienation and possible political defections, and legitimised post-War international governance. Hence development-as-security was long premised on establishing and maintaining minimal stability within a geopolitically defined commons, and it continues to be so. Although security remains a core development *dispositif*, humanitarian and development agencies have increasingly coordinated with political and military stakeholders in the post-Cold War period in order to ensure containment of the (Western) commons against global threats that include spontaneous migration, terrorism, and political-economically destabilising political dissent (Duffield 2007: 55; Roberts 2014). To head off such threats, national and international bodies put in place blocks: 'exclusionary walls rather than developmental stairways' in James Ferguson's evocative phrase (Ferguson 2005: 177). Unravelling the genealogies of development-as-security vividly demonstrates how counter-insurgency mindsets suture humanitarian interventions to military strategy (Duffield 2010: 60–2), modernity (Watts 1995), and coloniality (Maldonado-Torres 2007). Accordingly, more than 85 per cent of the major conflicts since the Second World War have been in poor countries (Meekosha 2011), a legacy contributing to high mortality as well as the after-effects of physical impairment and psychic trauma. Also appallingly numerous are the conflicts that cannot even be named as war although they impact on a daily basis on livelihoods, social relations, embodiments, and dispositions.[8]

Despite the important case made for basic needs, poverty-alleviation, and human rights, a global development commons neither equals post-War Keynesian welfare nor the ambitions of the mid-twentieth-century United Nations. Instead a now extensive literature speaks of how prevalent macroeconomic practice and policy has persistently resulted in direct, often violently enforced, forms of dispossession against which gestures towards a common humanity held little sway in a consistent or even manner. Everyday lives have become variously enmeshed in multiple processes of dispossession associated with mineral extraction, the deepening of markets and encroachment on to the natural world, an active disinvestment in public services, and the geographical extension and social locking-in of spaces of abandonment. In this sense, the terrain of majority-world/minority-world dynamics can also be viewed as processes of abandonment and the production of bare life (Scheper-Hughes 1992; Li 2010; Sylvester 2006). Outside malls and gated communities, the global South arguably becomes the site where unemployed or underemployed populations experience lives of quiet desperation. Humanitarian intervention in this context hence occurs after sudden collapses of over-stretched minimal public and private social facilities, together with overwhelming pressures on the for-the-most-part self-reliant impoverished populations (Duffield 2007). Attention has recently been focused on how the populations who bear the brunt of such abandonment, violence, and bare life are racialised, in ways that pump-prime different rules of engagement from those associated with white privilege (McIntyre and Nast 2011; for example, Haiti).

Across this wasted landscape moreover, countries in the global South and beyond have become the testing ground for knowledge production. South-based medical trials are generally found to contribute to uneven protection, ethics, and rights for the purported 'common' humanity. In such a biopolitics, the global South increasingly functions as a site for experimentation on bare life. Clinical trials have decisively moved outside the USA and Western Europe into developing countries as well as Eastern Europe and Russia (Glickman *et al.* 2009). Between 1995 and 2005, medical researchers found that the number of trial sites outside the United States had more than doubled, while the proportion of US-based and Western Europe-based trials had fallen. The relocation of trials to India and South America is driven by huge labour cost differentials, quicker processing times, and laxer regulatory frameworks leading to cost savings of around 90 per cent (Glickman *et al.* 2009; Petryna 2009).

The lopsided global dynamics of development are also bound up with the globally uneven distribution of different generations, as nine out of every ten children (87 per cent) are found in developing countries (Ruddick 2003: 340). Despite the Millennium Development Goals' undoubted impact on raising access to formal education and reducing infant mortality rates, the lack of a development commons is brought into focus by generation-specific questions around (lack of) employment, crumbling public services (particularly in education), forms of citizenship, and majority world young peoples' limited leverage vis-à-vis the North Atlantic. In areas affected by HIV-AIDS, for instance, orphaned young people increasingly rely on older generations whose own capacity to cope is under extreme pressure given the inadequate state of social and health services. Sharp increases in poverty among the majority world elderly speak vividly of the selective parameters of a global 'commons'.[9] Social programmes for young people's formal education and healthy life have, as yet, not evidenced long-term effects and 'the higher employment rates and better qualifications predicted among [participating poor] children ... remain largely within the realm of political conjecture' (Merrien 2013: 104; see also Section II). Economic restructuring has similarly entrenched child labour, involving around eighty million children worldwide, while fortress migration policies in wealthy countries particularly affect younger cohorts (Duffield 2007: Ruddick 2003). Despite the geopolitics of containment however, remittances comprise a rapidly growing component of financial flows, providing essential resources directly to some of the world's low-income individuals; global estimates suggest that in 2012 remittances equated to some US$401bn, equivalent to around one-third of OECD ODA. Remittances are highly skewed towards India, China, Mexico, Philippines, Egypt, and Pakistan, whereas sub-Saharan Africa generally accounts for small amounts.[10]

Furthermore, an estimated 80 per cent of the world's more than a billion disabled people live in the global South, the result of the *production* of impairment rather than the outcome of random chance (Chouinard 2014; Meekosha 2011). In the lopsided terrain of global inequality, people in the global South are disproportionately exposed to the processes that result in impairment and where the resources, knowledge, and political will to address their concerns are by and large missing. The global South becomes the site of disproportionate impairment because of its concentration of poorly-regulated work places which contribute to injury, impairment, and death, while the application of structural adjustment programmes curtail public health and social services, and contribute to the outmigration of health professionals to wealthy countries. In the meantime, because NGOs are only unevenly distributed they cannot make a systematic difference. In parallel, metropolitan countries dispose of waste in ways that compound impoverishment and differential valuation of human life, a form of environmental racism that shunts health- and environmentally-damaging waste onto racialised areas and inhabitants construed as of lesser worth (Pulido 2000).

In this context of contradictory material disengagement at a number of institutional scales, how does the fiction of a common humanity as an affective field of action get reproduced? Research has begun to unpack how common humanity is construed as an affect, a set of communicative and social-relation-building processes across individual and collective subjects which gains purchase in metropolitan political, public, and policy domains, despite the countervailing material processes described above. Affect undoubtedly informs emotional connections between Northern and Southern subjects yet 'these collectivities are differently capable of affecting and being affected because of their access to social/geopolitical power...' (Tolia-Kelly 2006: 215). Personal charitable donations to the majority world tend for instance to reproduce relational distinctions between the donor (as agent) and the recipient as vulnerable. As Ahmed notes, 'the over-representation of the pain of others is significant as it fixes the other as the one who "has" pain, and who can overcome that pain only when the western subject feels moved enough to give' (Ahmed 2004: 35; see also Sontag 2003). Donor common 'feeling' with donees is temporary, cited at the moment it is obviated by a donation, leaving in place enduring distinctions between different kinds of 'humanity'. Despite gesturing towards cosmopolitanism, such affective dynamics require no critical reflection on the relative positionality and structural conditions of giving, in terms of post-colonial political economies and subjectivities (Brennan 1997; Ahmed 2004).

This section has aimed to provide a panorama of the ways in which business as usual combined with material and affective disengagement, premised upon the selectivity and conditionality of assistance to the world's poor, have but scratched the surface of how to make a common humanity live up both to its ideals of generosity and appropriate response to need and their affective correlates.

II Helping the world's poor? Social protection in the twenty-first century

One significant sphere where political, economic, and affective-performative dimensions of what I have termed the commons-impoverishment paradox have been played out over the past two decades is social protection policy, a globally prominent paradigm ever since the Millennium Development Goals prioritised poverty alleviation. As James Ferguson notes, 'distributive outcomes for those at the bottom of the economic heap are increasingly determined within the domain of social policy' (Ferguson 2015: 24). Aiming to reverse processes of impoverishment in the wake of neoliberal restructuring and retrenchment, social protection calls up notions of community and the social to offer a more inclusive capitalism for the poorest of the poor, features particularly relevant for our analysis here (cf. Lemke 2001; Rose and Miller 2008; Lazarus 2008).[11] Retreating from earlier unmediated neoliberalism, social protection came to the fore under the post-Washington consensus[12] goal of development 'with a human face'. State cutbacks, labour market liberalisation, and asset privatisation were all set to continue but they were balanced by measures to mitigate their disastrous impact on the poorest. The UN's 1995 World Summit for Social Development marked the agenda's arrival, followed by the creation of Ministries of Social Development across the majority world (Molyneux 2008; Fakuda-Parr and Hulme 2011). Under social neoliberalism, programmes target interventions at specific subpopulations known through an association with vulnerability and risks, to furnish them with the capacity to better face neoliberal economic challenges (Best 2013; Merrien 2013). Social, economic, political, and environmental risks are foregrounded (Merrien 2013: 96), prompting development to deal with emergent volatility, complexity, uncertainty, and ambiguity through innovation and constant adaptation (e.g. Ramalingham 2013; for critiques, see Duffield 2010; Radcliffe 2016). According to global development agencies, social policies are not to be confused with universal welfare expenditure, as they comprise a means of strengthening social and human capital, whereby poor populations are to be helped to strategically deploy their personal and community-based networks and capacities to consolidate meagre levels of human capital, as measured by formal education, health, and the like (Holzmann and Jørgensen 2000; Merrien 2013). In turn, the poor are to become more 'resilient', a concept adapted from ecological studies, an aim that 'seeks to enhance an individual's or system's capacity to live with, or indeed prosper from, uncertainty' (Walker and Cooper 2011: 153). Nurturing resilience in Jamaica, for example, involves training low-income, marginal, urban populations to prepare for disasters by coordinating locally and informing themselves of privately-held resources to call upon in an emergency in a self-sufficient way (Grove 2014).

In other words, the provision of resources to the global poor is framed strongly through discourses around the need to target particular social categories, in order to address individual capacities to survive and thrive in what remains a vigorously market driven political economy. Following this logic ethnographically ('the social life of cash payments' according to Ferguson 2015: 136), reveals how implementing social protection relies on ever-finer and more pernicious distinctions between deserving beneficiaries and non-beneficiaries, grounded on contradictory interpretations of vulnerability *and* of the poor's potential to thrive under volatile capitalism. Categorising subjects and fine-tuning dispositions deflects attention away from ongoing structural shifts that defund public resources, validate regressive tax structures, and continue to treat the poor as unknowledgeable, incompetent actors, all of which show up the fiction of a 'development commons'.[13] Nevertheless, persistent impoverishment and eroding substantive citizenship had unintended transformational consequences leading to re-commoning politics, as will be traced in Section III.

II.1   Social protection's poster child: conditional cash transfers

Conditional cash transfers or CCTs comprise a rapidly-proliferating (probably *the* most rapidly extending) policy model used by governments and international agencies alike.[14] CCTs supply monetary grants and in some cases, food vouchers (in Latin America, CCTs average US$35 per month) to adult beneficiaries (over-whelmingly women, who are mothers or expectant mothers) on condition that recipients fulfil a number of duties including regular health checks and ensuring children attend school. Transfers represent the 'burden of securing one's liveli-hood ... squarely placed on the shoulders of individual households' (Arora and Romijn 2011: 482). Additionally, some CCTs require participants to volunteer for community activities, others for young children to attend pre-school, and others to enrol on welfare-for-work systems.

Starting in 1996 in Brazil and 1997 in Mexico, the oft-stated objective of CCTs is to permit low-income households to raise their human capital, namely the command over resources possible with more educated and healthier house-hold members. Changing conceptions of poverty's causes entail changing prac-tices, so that 'the techniques that manage it must also be more flexible and proactive' (Best 2013: 110). A range of methods are used to identify the CCTs' target group including demand-led registration, geographic targeting, means testing, proxy-means testing, community-based selection, while frequently more than one method is used (Merrien 2013). Major Latin American and Caribbean countries provide cash transfers under these conditions to over 20 per cent of their populations.[15] As a non-universal model, the CCTs rely on the establish-ment of processes, divisions of labour, and forms of interaction with beneficiar-ies to regularly check participants fulfil the criteria for assistance which entails bureaucratically calibrated categories and measures of geographical location, age, childcare responsibilities, and level of impoverishment. Peru's CCT pro-gramme known as 'Juntos' works in rural areas with impoverished households, deliberately excluding households with incomes derived from mining. Juntos is managed largely through decentralised local managers who check compliance with conditions by working closely with beneficiary women, triangulating data, and adjusting lists of eligible recipients and at times relying on public validation (Cookson 2015; Meltzer 2013: 643).[16]

In rural Peru public services are scarce, so women have to walk for hours to wait in line for confirmation that they meet the conditions of the transfer, as well as attend meetings called by local managers, to get to clinics for appointments, and to take children to school. Additional to the long hours of productive, repro-ductive, and community work of a low-income female, these tasks are associated with fulfilling a signed contract with the state. The lack of public transport infra-structures that might connect spaces of hyper-marginality and racialised internal colonies with the always-already better endowed spaces where money is disbursed, reflect and reproduce the value hierarchies of different groups of bodies (as I discuss further below). Such plays of power and difference also come to operate at the heart of Juntos' daily interactions between recipients and functionaries. In addition to formal conditions, female participants are induced into believing that other conditions apply, although in fact they arise from project employees' reliance on teachers, health workers and municipal officer to confirm data management, and thereby guarantee performance-related pay. Juntos local functionaries establish these shadow conditionalities (for example, eating protein, enrolling children in *pre*-school, attending meetings, and giving birth in a health centre, see Cookson 2015). Far from the components sanctioned at the level of Lima and Washington, shadow conditionalities comprise the messy social – but far from 'commoning' – outcome of neoliberal labour and institutionalisation re-tooled for distribution, and of established power hierarchies between urban and rural, professional and smallholder farmer, white and racialised subjects. Viewed from below, conditional cash transfers appear less as neutral largesse than as a means to make already difficult lives yet more enmeshed in opportunity- and energy-sapping obligations to the already-powerful. As Peruvian rural women's critiques make abundantly clear, this is not about a new 'commons' but rather the urgent need to attend poor quality, understaffed public services (again, see below). In place of mutual recognition and justice, the quotidian interactions underpinning Juntos produce beneficiaries' resentment, fear, and anger.

The impacts of micro-level social distinctions from positions of (relative) power are equally documented for Bangladesh's CCT programme, designed to guarantee poor children's attendance at school (Hossain 2010). There, CCT local staff continuously and silently make social judgements about potential beneficiaries, calling upon subtle distinctions of class to allocate finite resources between those perceived as 'deserving', impoverished middle-class subjects, and those tagged by contrast as 'undeserving'. Local workers hold socially-weighty judgements about (child) labour, caste-class and status, and make 'cost-benefit' calculations about which kids will give the greatest 'return' on transfers. Whereas Lagarde and the IMF express their visions of a singular, homogeneous, global poor, cash transfers on the ground rely upon historically and geographically entrenched socio-economic rationalities in which fine-grained, significant meanings around social difference are stickily attached to certain social labels. Such pre-existing meanings remain unexamined in projects, yet shape how beneficiaries are selected and the expectations placed on them, thereby providing some 'machines of morality' (Rose 1999; see, for example, Meltzer 2013).

Bridging the contradiction between dominant visions of 'poor humanity' and the reality of social distinctions is the largely unacknowledged but nonetheless critical understanding of social policy as gendered. As scholars have demonstrated, CCTs are highly gendered interventions as they rely upon hegemonic conceptions of female adults at the core of a domesticated sphere and masculine unreliability (Molyneux 2008). Moreover, the goal of building a younger generation's capacity relies upon assumptions that female subjects are available for the hard slog of building social relations and sacrificing individual advancement in the name of a postponed – albeit familial – good (Federici 2010). Gendered norms work through into shadow moral pressures too. In Uruguay, to take one example, urban neighbours distinguished between worthy female CCT recipients (those who set up 'appropriate' micro-enterprises) while condemning others (Corboz 2013: 77). Cash transfers also reinforce gendered distinctions between male breadwinners and feminised subjects who cannot be expected to work full time (Ferguson 2015).

CCTs' contradictory relationships with the global commons in its *diversity* can be unpacked further at a variety of scales. Many Latin America CCTs are, for instance, effectively *competitors* with common resources. Recipient households in Peru and Uruguay are required to attend crumbling public institutions or turn to market providers of services. As Latin America's CCT spending rose as a share of social budgets – with World Bank endorsement – so investment in public services such as health, education, and housing *fell* precipitously. Compared with income transfers that grew at 3.5 per cent between 1990–01 and 2008–09, spending on health rose by 1 per cent, and housing by a mere 0.4 per cent (Lavinas 2013: 20). Meanwhile, revenues dedicated to CCTs frequently rely on regressive sales tax. Providing the poor with cash 'rather than decommodified public goods or services' (Lavinas 2013: 7) hence becomes more consistent with global fictions of a common humanity than with a commons arising out of equal rights and substantive citizenship, providing ballast to criticisms that CCTs modestly reduce the rich–poor gap but do not deal with inequality. Nor does emptying out a public commons occur merely at the macroeconomic level. Ethnographic work on CCTs documents how local staff readily challenge recipients' claims on the state, referring to the need to break expectations of what staff term 'paternalism' (Meltzer 2013: 647; Cookson 2015). A citizen's right to make claims on the nation state is interpreted on-the-ground as a personal disposition at odds with a self-actualising, self-sufficient subjectivity implicit to social protection (on Africa, see Merrien 2013: 103).

II.II CCTs as financial instruments

Inclusive capitalism in the context of cash transfers hence emerges as a set of relations firmly embedded in capitalism as usual while drilling down into social distinctions between fiscally responsible and particularly-needy (and particularly gendered) subjects. Moreover, social orientations towards the nation state are recalibrated at a number of scales, while recipients and households remain enmeshed in prior hierarchical relationships with key gatekeepers to these meagre resources. These tendencies are consistent with a further, often underreported, dimension of cash transfers, namely their entanglement with processes of financialisation. Inclusive capitalism in this sense means *financial* inclusion, the enrolment of low-income individuals into ties with banking institutions. Social protection measures arguably construe a common humanity around economic transactions, an 'economic citizenship' firmly embedded within a landscape dominated by corporations and capital philanthropy. Assemblages of ICTs and financial instruments become the modality through which such inclusion is realized. Policy documents speak about the need to ensure 'financial inclusion' and asset-building, in order to build recipients' 'knowledge about money management... increasing self-esteem' (Meltzer 2013: 644). This section examines how these cash transfers engage recipients in such financialisation.

On the ground, CCTs now often include a component of training recipients in how to better manage financial assets and to use new communication technologies (Schwittay 2011). CCTs use a variety of private and state banking institutions, some of which are largely deregulated, even after the banking crisis. Peru's CCT programme was the first to directly incorporate a financial dimension through its Proyecto Capital (Project Capital), which aims to induct savings practices among beneficiaries through a system of incentives (which might include matching grants, subsidised interest rates, and food raffles). Despite fluently managing numerous monetary and non-monetary systems of exchange, valuation, and markets, low-income women are offered training workshops to teach them how to take out and manage loans (Lavinas 2013: 36).[17] Juntos financial workshops are provided by a lobbying group, a 'flex organisation' occupying a grey area between public programme and private gain to 'access resources and bypass conventional political constraints and forms of accountability' (Larner 2011: 329; cf. Roberts 2014). At the grass roots then, the financialisation agenda has ambivalent effects on local practices of commoning and the moulding of a genuinely *common* agenda.

Development agencies, including the Ford Foundation and the Inter-American Development Bank, financial corporations, and large companies have expressed great interest in this trend of social protection. They endorse Hernando de Soto's argument that building (legally titled) assets is the way to *immunise* the world's poor against the risk of falling ever deeper into poverty. Development agencies and philanthropic organisations invest time and money in refining measurements, specifying optimum project designs, and linking in with large-scale financial institutions. The Bill and Melinda Gates Foundation also supports initiatives with MasterCard in Africa (Lavinas 2013: 36), and finances the Global Financial Inclusion Database to generate data on individual saving and borrowing. Meanwhile the UK's Department for International Development runs a Financial Education Fund (Meltzer 2013: 644).[18] While such moves celebrate individual economic rights over public shared facilities, perhaps more important is the documented interest in accessing the latent purchasing power of the world's poorest in part through establishing technological reach (Arora and Romijn 2011; Ramalingham 2013). Such trends appear to be entirely consistent with certain private philanthropic initiatives operating through assemblages of mobile technology (for example, Black 2009).[19]

The NGO efforts to ensure no-strings-attached cash to low-income households in Niger, mentioned in the chapter's introduction, must thus be placed within a wider context (Hanlon *et al.* 2010). Unconditional cash transfers award money to selected groups without any conditions. In contexts of food production crisis, unconditional cash transfers are distributed to sustain local food markets following Sen's work on famine (Hanlon *et al.* 2010). In this profound reconfiguring of what a commons might even look like, certain voices are legitimate, others are not, resulting often from epistemic violence arising from post-colonial intersectional hierarchies (Arora and Romijn 2011; Radcliffe 2015). Routine development-beneficiary interfaces reproduce pejorative distinctions again and again, once more stickily attached to its subjects (Ahmed 2004), and to the interpersonal dynamics through which majority-world subjects are meant to access even these meagre resources. Whereas schemes offering unconditional transfers speak to the crucial recognition that the poor know best how to spend money appropriately (Ferguson 2015), the financialisation of social protection risks perpetuating the deeply engrained dynamics of post-colonial difference. The conception of a *developed* subject as the epitome of freedom has its roots in colonial-imperial dynamics of power that pre-date neoliberal doctrines by centuries. As post-colonial critique makes clear, the liberal category of the human has been identified and celebrated through opposition to nature, the subaltern, the racialised, and women. Development instrumentalisation of the social has thus to be placed within this analytical frame. Western concepts of liberty and freedom were established around the 'parvenu self-enclosed' subject, detached from social embeddedness, collectivity, and culture, and composed through relational oppositions with culturally different, unfree subject (Povinelli 2005). Becoming free, the subject accesses 'new institutions of risk and pleasure that make freedom from social relations seem natural and desirable' (Povinelli 2005: 162).

III Post-neoliberal promise and the contested process of constructing a (problematic) commons

In the face of the global trends in inequality and erosion of common purpose, the dispossessed in Latin America have nevertheless bitten back by electing post-neoliberal governments charged with moral responsibility for delivering citizens' inalienable rights and building particular commons (Grugel and Riggirozzi 2012). Placing the state at the heart of developmental agendas, post-neoliberal governments emerged on waves of popular mobilisations that have led to new rights regimes becoming codified in constitutions, often after lengthy consultation and public debate, in part based upon new publics founded on the streets and neighbourhoods. As the case of Ecuador illustrates, these regimes then implement agendas based on place-specific mixes of socialism, indigenous principles, and renegotiated resource-extraction contracts (Grugel and Riggirozzi 2012; Radcliffe 2012). While not overthrowing linkages into global capitalist economies, these new rights regimes have responded to the votes of the dispossessed and impoverished middle classes alike, generating complex processes of disarticulating – if admittedly not dismantling – neoliberalism (Escobar 2010; Grugel and Riggirozzi 2012). At the same time, the post-neoliberal state's largesse often depends on exporting natural and mineral resources, leaving in place unequal divisions of labour and threatening the global environmental commons. Under the umbrella of governmental experimentation, intellectuals and civil society actively re-imagine a new commons whether as a form of governance or as principles for a reformed social contract (El Bien Común 2013; Tapia 2006).[20]

Such oppositional politics often come to turn on the rejection of principles of privatisation and restructuring that held sway in Latin America from the 1970s on, leading in the early to mid-2000s to variable combinations of export taxes, renegotiated contracts with transnational petrol and mineral companies, and the search for non-hegemonic trading and geopolitical allies (including China). With booming primary commodity prices, post-neoliberal governments have had the fiscal resources with which to transform social protection within the context of resource nationalism and populist expressions of sovereignty. Ecuador vividly exemplifies these dynamics, inaugurating new constitutional, legislative, and governance principles for this oil-dependent middle-income but highly unequal country. After a lengthy consultative process, the 2008 Constitution was ratified massively by popular vote and the government became committed to the realization of an ambitious programme of rights-based development and macroeconomic growth oriented towards citizen well-being. The renegotiation of oil contracts released resources for social programmes. According to the latest development plan, preference has been given to resource distribution resulting in Latin America's greatest decline in inequality 2007–11. 'Sowing petrol' is particularly impactful in social programmes, where spending has soared. Efforts have also been made to rebuild public space as a commons, with emphasis placed on strengthening national identity, recognising diverse identities, and addressing gender-based violence (Ecuador 2013: 17).

These issues particularly come to the fore as Ecuador's social welfare spending has risen sharply, to historically unprecedented levels although at lower than worldwide averages. Public education is now free of charge up to university level, as are medical consultations. Nearly six out of every ten pensioners now have non-contributory pensions while cash transfers remain conditional although they have risen in number, amount, and coverage (consideration is now being given to vary amounts awarded by ethnicity, children's age, area, and income quintile). Currently, around 44 per cent of the population receives a cash transfer of US$30–50. By 2011, 9.3 per cent of GDP was spent on public and social measures of these kinds. Development plans forecast moves towards a non-contributory 'universal social protection floor' (Ecuador 2013: 119). Funded primarily through banking transaction charges and mineral exports, Ecuador's social spending however appears rather less like structural redistribution, and more like previous eras. Conditionality remains firmly in place for sure; recent policy debates propose that conditions be better policed than previously and mothers be required to receive more training. Despite infrastructure investment, public services also remain uneven in quality and presence, resulting in the slow amplification of 'the frontier of citizenship' (Radcliffe 2015). Unlike southern Africa say, where principles of fair share animate discussions of an unconditional basic income grant (BIG) to all citizens (Ballard 2013; Ferguson 2015), the Ecuadorian case suggests that the system of providing for the poor remains, however implicitly, embedded within the reproduction of inequality.

Post-neoliberal Ecuador's understandable commitment to the (re)construction of a national commons premised upon equal rights among unequally endowed citizens nevertheless raises profound questions about the longer-term viability of the model and its capacity to build a 'commons'. China's faltering growth and falling world oil prices put at risk the fiscal sustainability of social welfare spending. Equally contentious are the consequences of the 'commodities consensus' (Svampa 2015) for the global environment, the fragile commons of the Anthropocene. While at one time, Ecuador's president pledged to keep oil underground in return for international donations to pay for biodiversity protection in the Yasuní region, that scheme is no longer on the table; indeed, mineral extraction accelerated during his second term. Gustavo Esteva's comment that the 'commons is not a universal category' (Esteva 2015: 3) pinpoints precisely what is at stake in Ecuador, where new alliances of third sector and civil society groups mobilise to protect what they consider their rightful commons, including water, heaths, landholding communities, and indigenous life-ways.

IV Conclusions

Global inequality and poverty represent a challenging arena within which to think through the commons and its practices, its exclusions and points of expansion. Given the scale of the issues and the overlapping economic, social, affective, and political dimensions to the dilemma of how to think about – let alone act upon – the gross and grotesque disparities characteristic of the world today is almost impossible. Following one thread through that complexity, this chapter has pondered on what it means to think about a 'common humanity' that elicits a response of responsibility and common identification. Gestures of Northern concern and generosity occur alongside – and through – the protracted degrading of peoples' lives. Such accounts have been written before. What I have found distinctive however about today's commons-impoverishment paradox, examined ethnographically, are the ways in which a 'common humanity' is quickly subdivided and distinctions are made around affective-performative power and governmental designations of subjects. While such an account echoes Foucauldian critiques of governmentality, it also reveals how colonial-modern distinctions and hierarchies animate neoliberal policies, silently working to build up ontological, epistemological, and social barriers between global South and North. Financialisation perhaps indicates a different bridging of the fault-lines in a 'common humanity', dispensing as it does with interpersonal interactions in favour of technologically-mediated flows in a distributed economically-based network. Each of these processes of course remains contested and subject to reinterpretation and re-theorisation whether in the streets or the fields. While humanitarian projects may always be accompanied by a specific 'moral vernacular' (Lester and Dussart 2012: 2), the contemporary disparity between a restricted, conditional gesture towards the poor and 'inclusive capitalism's' expansive claims raises urgent questions about where to go next. James Ferguson concludes that a new politics of distribution is discernible in debates around *un*conditional cash transfers, the hopeful horizon of a commons premised on a person's 'rightful share' which 'has no expectations of a return, no debt and no shame' (Ferguson 2015: 178). Exploring these debates in Latin America by contrast suggests that a progressive politics of redistribution relies upon constitutional and legislative transformation, in addition to shifts in tax and economic structures, as well as a deep rethinking of the social hierarchies that characterise the region. Ferguson's endorsement of a basic income grant premised on universal access, with no social judgement – no moral vernacular – concerning the recipient's social, familial, or employment qualities, does indeed go a long way towards suturing justice to a common humanity's claims on resources. Yet as long as social distinctions, grounded in lopsided post-colonial hierarchies of social difference, remain the core driver and framework for poverty alleviation efforts, as they so frequently do often in unacknowledged ways, the commons-impoverishment paradox will remain firmly in place.

Notes

1 I am grateful to David Nally, Emma Mawdsley, Marilyn Strathern, Ash Amin, and Symposium participants for comments on an earlier version; any errors remain mine.
2 Development refers here to immanent processes of economic growth, and specific programmes by nation states, multilateral and bilateral agencies, non-governmental organisations (NGOs), and civil society organisations.
3 I focus on Northern Atlantic institutions and their global impacts; cf. Mawdsley 2012.
4 According to the World Bank, worldwide remittances to developing countries reached their highest level in 2012 at US$389 billion, ahead of ODA and following FDI.
5 OECD press release 'Development: Aid to Developing Countries Falls because of Global Recession', issue 4, April 2012.
6 Aid's switch to Europe had a higher standard of 'emergency' infrastructure, suggesting that development deals with geographically variable 'kinds of humanity', some worthy of comfort and dignity.
7 The UK's Department for International Development gave to seventy-eight countries in 2010–11, and to twenty-eight countries by 2012.
8 Mexico's 'war on drugs' has killed or injured around 120,000 people and led to 27,000 missing persons over the past decade.
9 Meanwhile minority-world elderly are cared for by global immigrants (e.g. South Americans in Spain and Italy).
10 Nigeria is sub-Saharan Africa's largest remittance recipient (with 65 per cent of officially-recorded regional remittance flows and 2 per cent of global flows).
11 A genealogy of social protection is historically and geographically contingent (Merrien 2013; Ferguson 2015); I trace a Latin American-inflected genealogy with an inevitable 'view from somewhere'.
12 Post-Washington consensus refers to a hegemonic reorientation of neoliberalism arising after Structural Adjustment Policies and Latin America's 'lost decade' of the 1980s.
13 This interpretation runs up against mainstream evaluations of social protection, based overwhelmingly on quantitative survey data.

14 Social protection measures also include public and private social insurance for formal sector workers, and pensions (Ballard 2013; Merrien 2013; Barrientos *et al.* 2013).

15 Brazil (26 per cent of national population), Colombia (25 per cent), Mexico (25 per cent), Guatemala 23 per cent, Dominican Republic (21 per cent): see Lavinas 2013: 19.

16 In Latin America, CCT recipients are overwhelmingly mothers and mothers-to-be.

17 CCTs' financialisation largely ignores existing familiarity with financial institutions, such as remittance transfer via companies such as Western Union where users pay 13–20 per cent of money remitted on transaction costs, cutting benefit for families back home while boosting company profits.

18 The RAND Corporation currently supports a two-year Economic and Social Research Council (ESRC, UK) and UK Department for International Development Project on CCT effectiveness in four Latin American countries.

19 For instance,

> A satellite passing over east Africa took pictures of [roofs].... [M]onitoring the satellite data remotely [indicates roof material, a proxy for poverty] ... Google and other donors contributed money to ... a charity which hands out no-strings-attached cash to the poorest people it can find.
>
> ('Cash to the Poor: Pennies from Heaven', *The Economist*, 26 October 2013)

Poor households receive a (free) mobile phone, presumed to be the medium lacking in this context (cf. Ramalingham 2013).

20 The Common Good Pact is a network of organisations, academics, and NGOs lobbying for a fiscal agreement between state and society to ensure greater tax revenue and distribution.

References

Ahmed, Sara. 'Collective Feelings, Or The Impressions Left by Others'. *Theory, Culture and Society* 21, no. 2 (2004): 25–42.

Arora, Saurabh and Henny Romijn. 'The Empty Rhetoric of Poverty Reduction at the Base of the Pyramid'. *Organization* 19, no. 4 (2011): 481–505.

Ballard, Richard. 'Geographies of Development II: Cash Transfers and the Reinvention of Development for the Poor'. *Progress in Human Geography* 37, no. 6 (2013): 811–21.

Barrientos, Armando, Vera Møller, Joao Saboia, Peter Lloyd-Sherlock, and Julia Mase. ' "Growing" Social Protection in Developing Countries: Lessons from Brazil and South Africa'. *Development Southern Africa* 30, no. 1 (2013): 54–68.

Best, Jacqueline. 'Redefining Poverty as Risk and Vulnerability: Shifting Strategies of Liberal Economic Governance'. *Third World Quarterly* 34, no. 1 (2013): 109–29.

Black, Shameem. 'Microloans and Micronarratives: Sentiment for a Small World'. *Public Culture* 21, no. 2 (2009): 269–92.

Brennan, Timothy. *At Home in the World: Cosmopolitanism Now.* Cambridge, MA: Harvard University Press, 1997.

CaLP Case Study. *Unconditional Cash Transfers to Reduce Food Insecurity for Displaced Households and Assist in the Repatriation of People to their Villages of Origin: Regions of Zinder, Agadez and Maradi in Niger.* Oxford: Oxfam, n.d.

Chouinard, Vera. 'Precarious Lives in the Global South: On Being Disabled in Guyana'. *Antipode* 46, no. 2 (2014): 340–58.

Cookson, Tara. *Rural Women and the Uneven Process of Inclusion: An Institutional Ethnography of Peru's Conditional Cash Transfer Programme.* Unpublished PhD, Department of Geography, University of Cambridge, 2015.

Corboz, Julienne. 'Third-Way Neoliberalism and Conditional Cash Transfers: The Paradoxes of Empowerment, Participation and Self-Worth among Poor Uruguayan Women'. *Australian Journal of Anthropology* 24, no. 1 (2013): 64–80.

Duffield, Mark. *Development, Security and Unending War: Governing the World of Peoples*. Cambridge: Polity, 2007.

Duffield, Mark. 'The Liberal Way of Development and the Development-Security Impasse: Exploring the Global Life-Chance Divide'. *Security Dialogue* 41, no. 1 (2010): 53–76.

Duffield, Mark and Vernon Hewitt (eds). *Empire, Development and Colonialism: The Past in the Present*. London: James Currey, 2009.

Ecuador. *Buen Vivir: Plan Nacional* 2013–2017. Quito: SENPLADES, 2013.

El Bien Común. *Pacto por el Bien Común*. La Paz: El Bien Común, 2013.

Escobar, Arturo. 'Latin America at the Cross-Roads: Alternative Modernizations, Post-Liberalism or Post-Development?' *Cultural Studies* 24, no. 1 (2010): 1–65.

Esteva, Gustavo. 'Conversing on the Commons: an Interview with Gustavo Esteva – Part 2. Orla O'Donovan'. *Community Development Journal* (2015) (DOI: 10.1093/cdj/bsv014).

Fanon, Frantz. *The Wretched of the Earth*. London: Penguin, 2001 [1961].

Federici, Silvia. 'Feminism and the Politics of the Commons in an Era of Primitive Accumulation', in Craig Hughes, Stevie Peace, and Kevin Van Meter (eds) for the Team Colors Collective, *Uses of a Whirlwind: Movement, Movements, and Contemporary Radical Currents in the United States*. Oakland, CA: AK Press, 2010, 283–93; also at: www.commoner.org.uk/?p=113 (accessed 18 January 2016).

Ferguson, James. *The Anti-Politics Machine: 'Development', Depoliticization and Bureaucratic Power in Lesotho*. Cambridge: Cambridge University Press, 1990.

Ferguson, James. 'Decomposing Modernity: History and Hierarchy after Development', in Ania Loomba (ed.), *Postcolonial Studies and Beyond*. Durham, NC: Duke University Press, 2005, 166–81.

Ferguson, James. *Give a Man a Fish: Reflections on the New Politics of Distribution*. Durham, NC: Duke University Press, 2015.

Fukuda-Parr, Sakiko and David Hulme. 'International Norm Dynamics and the "End of Poverty": Understanding the Millennium Development Goals'. *Global Governance* 17 (2011): 17–36.

Glickman, Seth, John McHutchison, Eric Peterson, Charles B. Cairns, Robert A. Harrington, Robert M. Califf, and Kevin A. Schulman. 'Ethical and Scientific Implications of the Globalization of Clinical Research'. *New England Journal of Medicine* 360, no. 8 (2009): 816–23.

Grove, Kevin. 'Agency, Affect and the Immunological Politics of Disaster Resilience'. *Environment and Planning D: Society and Space* 32, no. 2 (2014): 240–56.

Grugel, Jean and Pía Riggirozzi. 'Post-Neoliberalism in Latin America: Rebuilding and Reclaiming the State after Crisis'. *Development and Change* 43, no. 1 (2012): 1–21.

Hanlon, Joseph, Armando Barrientos, and David Hulme. *Just Give Money to the Poor: The Development Revolution from the Global South*. Boulder, CO: Lynne Reinner, 2010.

Hermes, Niels and Robert Lensink. 'Changing the Conditions for Aid: A New Paradigm?' *Journal of Development Studies* 37, no. 6 (2001): 1–16.

Hickey, Sam and Sarah Bracking. 'Exploring the Politics of Chronic Poverty: From Representation to a Politics of Justice?' *World Development* 33, no. 6 (2005): 851–65.

Holzmann, Robert and Steen Jørgensen. *Social Risk Management: A New Conceptual Framework for Social Protection and Beyond*. Discussion Paper Series #6. Washington, DC: World Bank. 2000.

Hossain, Naomi. 'School Exclusion as Social Exclusion: The Practices and Effects of a Conditional Cash Transfer Programme for the Poor, Bangladesh'. *Journal of Development Studies* 46, no. 7 (2010): 1264–82.

Lagarde, Christine. *Economic Inclusion and Financial Integrity.* Address to the Conference on Inclusive Capitalism, London, 27 May 2014, at: www.imf.org.

Larner, Wendy. 'C-change? Geographies of Crisis'. *Dialogues in Human Geography* 1, no. 3 (2011): 319–35.

Lavinas, Lena. '21st Century Welfare'. *New Left Review* 84 (2013): 5–40.

Lazarus, J. 'Participation in Poverty Reduction Strategy Papers: Reviewing the Past, Assessing the Present and Predicting the Future'. *Third World Quarterly* 29, no. 6 (2008): 1205–21.

Lemke, Thomas. '"The Birth of Bio-Politics": Michel Foucault's Lecture at the Collège de France on Neoliberal Governmentality'. *Economy and Society* 30, no. 2 (2001): 190–207.

Lester, Alan and Fae Dussart. *Colonization and the Origins of Humanitarian Governance.* Cambridge: Cambridge University Press, 2012.

Li, Tania M. *The Will to Improve: Governmentality, Development and the Practice of Politics.* Durham, NC: Duke University Press, 2007.

Li, Tania M. 'To Make Live or Let Die? Rural Dispossession and the Protection of Surplus Populations'. *Antipode* 41, no. S1 (2010): 66–93.

Maldonado-Torres, N. 'On the Coloniality of Being: Contributions to the Development of a Concept.' *Cultural Studies* 21, no. 2–3 (2007): 240–70.

Mawdsley, Emma. *From Recipients to Donors: Emerging Powers and the Changing Development Landscape.* London: Zed, 2012.

McIntyre, Michael and Heidi Nast. 'Bio(necro)polis: Marx, Surplus Populations, and the Spatial Dialectics of Reproduction and "Race"'. *Antipode* 43, no. 5 (2011): 1465–88.

Meekosha, Helen. 'Decolonizing Disability: Thinking and Acting Globally'. *Disability and Society* 26, no. 6 (2011): 667–82.

Meltzer, Judy. '"Good Citizenship" and the Promotion of Personal Savings Accounts in Peru'. *Citizenship Studies* 17, no. 5 (2013): 641–52.

Merrien, François-Xavier. 'Social Protection as Development Policy: A New International Agenda for Action'. *International Development Policy* 5, no. 1 (2013): 89–106.

Molyneux, Maxine. 'The Neoliberal Turn and New Social Policy in Latin America'. *Development and Change* 39, no. 5 (2008): 775–97.

Petryna, Adriana. *When Experiments Travel: Clinical Trials and the Global Search for Human Subjects.* Princeton, NJ: Princeton University Press, 2009.

Povinelli, Elizabeth. 'A Flight from Freedom', in Ania Loomba (ed.), *Postcolonial Studies and Beyond.* Durham, NC: Duke University Press, 2005, 145–65.

Pulido, L. 'Rethinking Environmental Racism: White Privilege and Urban Development in Southern California.' *Annals of the American Association of Geographers* 90, no. 1 (2000): 12–40.

Radcliffe, Sarah A. 'Development for a Postneoliberal Era? Sumak Kawsay, Living Well and the Limits to Decolonization in Ecuador'. *Geoforum* 43, no. 2 (2012): 240–9.

Radcliffe, Sarah A. *Dilemmas of Difference: Indigenous Women and the Limits of Postcolonial Development Policy.* Durham, NC: Duke University Press, 2015.

Radcliffe, Sarah A. 'Civil Society: Management, Mismanagement and Informal Governance', in Daniel Hammett and Jean Grugel (eds), *Palgrave Handbook of International Development.* London: Palgrave, forthcoming, 2016.

Radcliffe, Sarah A. 'Género y Buen Vivir: Desigualdades Interseccionales y la Descolonización de las Jerarquías Persistentes', in Soledad Varea (ed.), *Feminismos y Buen Vivir.* Quito: Yachay Tech, forthcoming.

Ramalingham, Ben. *Aid on the Edge of Chaos.* Oxford: Oxford University Press, 2013.

Roberts, Susan. 'Development Capital: USAID and the Rise of Development Contractors'. *Annals of the Association of American Geographers* 104, no. 5 (2014): 1030–51.

Rose, Nikolas. *Powers of Freedom: Reframing Political Thought.* Cambridge: Cambridge University Press, 1999.

Rose, Nikolas and Peter Miller. *Governing the Present: Administering Economic, Social and Personal Life.* Cambridge: Polity, 2008.

Ruddick, Susan. 'The Politics of Aging: Globalization and the Restructuring of Youth and Childhood'. *Antipode* 35, no. 2 (2003): 334–62.

Scheper-Hughes, Nancy. *Death Without Weeping: The Violence of Everyday Life in Brazil.* Berkeley, CA: University of California Press, 1992.

Schwittay, Anke. 'The Financial Inclusion Assemblage: Subjects, Technics, Rationalities'. *Critique of Anthropology* 31, no. 4 (2011): 381–401.

Sontag, S. *Regarding the Pain of Others.* New York: Picador, 2003.

Svampa, Maristella. 'Commodities Consensus: Neoextractivism and Enclosure of the Commons in Latin America'. *South Atlantic Quarterly* 114, no. 1 (2015): 65–82.

Sylvester, Christine. 'Bare Life as a Development/Postcolonial Problematic'. *Geographical Journal* 172, no. 1 (2006): 66–77.

Tapia, Luis. *La Invención del Núcleo Común.* La Paz: Muela del Diablo, 2006.

Tolia-Kelly, Divya. 'Affect: an Ethnocentric Encounter? Exploring the "Universalist" Imperative of Emotional/Affectual Geographies'. *Area* 38, no. 2 (2006): 213–17.

Walker, Jeremy and Melinda Cooper. 'Genealogies of Resilience: From Systems Ecology to the Political Economy of Crisis Adaptation'. *Security Dialogue* 42, no. 2 (2011): 143–60.

Watts, M.J. 'A New Deal in Emotions', in J. Crush (ed.), *Power of Development.* London: Routledge, 1995, 44–62.

White, Howard and Simon Feeny. 'An Examination of the Long-Run Trends and Recent Developments in Foreign Aid'. *Journal of Economic Development* 28, no. 1 (2003): 113–35.

9 The urban metabolic commons

Rights, civil society, and subaltern struggle[1]

Colin McFarlane and Renu Desai

Introduction

The chapter focuses on the 'metabolic commons' of water and sanitation. Our concern is both with the ways in which these commons are siphoned off – leaving the urban poor struggling to meet basic needs of hydration or to access a clean, safe place to relieve themselves – and with political struggles for a more socially just commons. This is the commons of basic staples. We will examine three imaginaries of the urban metabolic commons: *rights-based civil society*, *entrepreneurial civil society*, and *subaltern politics*, each of which differently conceptualise the urban political and the actors that constitute it. Across the three instances there is a shared commitment to reducing poverty, oppression, and exploitation, to a better distribution of resource and opportunity, and to 'protecting' and 'prospecting' that which is broadly understood as being in common (Amin and Howell: Chapter 1).

However, there are important differences between the three forms in how political struggle is understood and practised, and on the role of different groups such as the state, activists, residents, and so on within that struggle. Rather than argue in favour of any one of these imaginaries as being more effective in mapping out the urban commons over any of the others, we instead treat them as 'commoning experiments' that reflect the non-singular nature of the commons. These experiments, often carried out in extremely difficult and precarious circumstances – ensuring access to a toilet, for example – illustrate the need for a generative conception of the commons whereby different 'arts of the political' (Amin and Thrift 2013) work more or less effectively in different contexts.

The commons refers to a wide set of more or less visible ways in which urban resources are organised, distributed, and policed (Gidwani and Baviskar 2011; Jeffrey *et al.* 2012; Parthasarathy 2011). We take the commons to signal a more expansive set of geographies than simply 'public space'. Water and sanitation relate, of course, to public space in all sorts of ways – for example, through the connections between public toilets or water standpipes and community dynamics – but they are also much more. Water and sanitation are produced through forms of infrastructural provisioning and curtailment, reflected in the differentiated geographies of pipes, toilets, and drains, and are governed through formal and informal regulations in often overlapping ways, especially in relation to informal settlements.

Urban metabolic perspectives have traced urbanisation and inequality in flows and internments organised through social and biophysical networks, including bodies, infrastructures, political economies, and services (e.g. Gandy 2004; Loftus 2009; Swyngedouw 2004). Gandy (2004) places an important emphasis on urban metabolic *transformation*, where metabolism refers not to anatomical or functionalist perspectives in a self-regulatory system, but to unequal political reconfigurations of nature and bodies in different conditions of capitalist urbanisation.

The idea of metabolism is a powerful way to interrogate the inequalities of the urban commons. Metabolism refers to bodily processes through which resources are used, transformed, put into circulation, and turned into energy or waste. To shift the notion of metabolism from the body to the urban retains a focus on the body, meaning that issues of hydration and waste remain crucial, but forces us to ask: what are the processes that shape the bodily urban experience? This requires revealing the ways in which water and waste are used, distributed, processed, and transformed through not just bodies, but through infrastructures, materials of different sorts (e.g. water buckets and improvised latrines), state policies, informal power brokers, cultures of control and exclusion, and economic processes such as costings or privatisation. These processes constitute the ways in which water and waste are differently metabolised in the city, with severe consequences for the urban poor, positioning the *metabolic lens* as a powerful tool for revealing the profound inequalities of the urban commons.

Our reflections on the urban metabolic commons and their politicisation are drawn from Mumbai, but we use this case to make a larger argument: that in evaluating the stakes and possibilities of different versions of the commons, we place an emphasis on experimentation and possibility over forms of judgement that dismiss this or that version of the commons and endorse another. The inequalities of power between the dominant political and economic actors and processes in contemporary cities and the struggles of the dispossessed and oppressed require multiple, often creative strategies. Critical academic and political practice is more likely to succeed if it works through and nurtures this generative multiplicity. We will argue for a particular sensibility of critique, one that aims to proliferate and assemble different forms of commoning, that evaluates their respective stakes but that values the multiplicity of the commons as itself an ethic of commoning that is in need of support.

For an increasing number of urban residents in India, the struggle for the urban commons is a metabolic struggle. There are many examples we could point to here, but two instances, both in relation to Mumbai's water and sanitation geographies, help to illustrate how the authorities – the municipal and regional state in these cases – make metabolic choices that too often entrench rather than alleviate the metabolic struggles of urban residents.

First, in 2010, following a poor monsoon and subsequent debates about 'water shortages' among the middle classes, the municipality launched a crackdown on 'illegal' water connections in informal settlements. Neighbourhoods that were 'illegal' and that lacked reliable political connections stood a good chance of having their water violently removed as pipes were cut and motors were seized. The removal of water by the municipality, often without notice, from informal neighbourhoods in Mumbai is not itself new, but in the backdrop of middle-class anxieties about 'shortages', it took on a new scale and intensity and was aided by the police.

But while this was going on in some of the most deprived areas of the city, especially in north-east Mumbai, the state was continuing to provide huge quantities of water at favourably low prices to the booming bottled drinks industry, one of the fastest growing and globalised industries in India, including dominant players like Pepsi, Nestlé, and Coca-Cola. To feed this resource-intensive industry, water is channelled through rural Maharashtra into Mumbai, bypassing en route not only the informal neighbourhoods violently denied a fundamental provision but also subsistence farmers whose poorly irrigated fields continue to fail while powerful sugar barons and commercial flower growers sap water (Graham *et al.* 2013).

The second example is earlier, from 2006, when Mumbai saw the legislation of the *Greater Mumbai Cleanliness and Sanitation Bye-laws* that introduced punitive measures against cooking, bathing, spitting, urinating, and defecating in public spaces. These by-laws – which regulate a variety of other activities like littering, waste segregation, and so on – are aimed at disciplining all urban residents and elevating that most politicised of urban discourses in India, 'civic consciousness', many of the punitive measures based more accurately on what Amita Baviskar (2002) refers to as 'bourgeois environmentalism'. Increasingly central to the politics of urban India's metabolic modernity, especially in the country's megacities, bourgeois environmentalism casts middle-class concerns around aesthetics, leisure, and health – often at the cost of informal settlements or street hawkers – as seemingly class-neutral discourses of the environmental quality of life.

In introducing disciplinary action against open defecation in a city in which around 25 per cent of residents have no or inadequate sanitation facilities (MW-YUVA 2001: 10), the by-laws positioned fundamental bodily needs against the right to a clean and sanitary environment (Desai *et al.* 2015). Here, discipline carries its familiar double meaning in relation to waste (Gidwani 2013): to remove waste from the urban environment, and to bring order to those deemed wasteful and indigent by the powerful minority, and who don't qualify as a visible part of India's urban metabolic modernity.

In the next section we set out some of the context of sanitation inequalities in India, with a particular focus on Mumbai's informal settlements. We then move on in the main body of the chapter to discuss three imaginaries of the urban commons: rights-based civil society, entrepreneurial civil society, and subaltern politics. We conclude by reflecting on the prospects for a more just urban metabolic commons in Mumbai and beyond.

Violence, the body, and metabolising waste

It is difficult to imagine a more profound illustration of the necessity of sanitation to life itself: two teenage girls venturing into the fields at night, are brutally raped, killed and left to hang from a mango tree. They left their homes because they had no alternative, due to the denial of adequate sanitation, but to answer the call of nature by use of a nearby field. This is a shocking and extremely sad story of violence and vulnerability in rural Uttar Pradesh, India, in 2014. A vulnerability produced by caste oppression, the normalisation of horrific violence against poor women, and the structural and systematic failure of the Indian state to provide the minimum of everyday rights: a clean, functional toilet. The *National Campaign on Dalit Human Rights* reported that 67 per cent of low-caste Dalit women – often referred to as 'untouchable' – in India have faced some form of sexual violence (Soundararajan 2014). This is likely to be an underestimate. Many of these attacks happen because women are forced into open defecation in fields, railway tracks, forest areas, garbage grounds, and other marginal spaces across India. The denial of adequate toilets, a profound and fundamental bodily need, gives rise to an opportunity for the worst kind of bodily violence. Partly as a result of such horrifying attacks, the question of open defecation has been placed firmly on the national agenda and has been taken up in public pronouncements by Prime Minister Narendra Modi to provide all Indian homes with toilets and eliminate open defecation by 2019.

Practices of open defecation deepen social inequalities in various ways, especially among impoverished women and children. In Mumbai, non-governmental organisations (NGOs) working in Rafinagar, a non-notified or 'illegal' informal settlement in the historically very poor north-east of the city, have noted the high incidence of diarrhoea, dysentery, and worms. Rafinagar is part of the city's 'slum belt' or 'malnutrition belt', one of the densest areas and described in the 2009 Mumbai Human Development Report as having the lowest scores for human development and the highest rates of infant mortality. Despite Mumbai's status as the wealthiest city in India, the report notes that in recent years, 'if anything has changed, it is the deterioration in health and sanitation conditions and the increasing social trauma of visible inequity' (Mumbai HDR 2010). As a network process, it is impossible to disentangle sanitation from other metabolic struggles, especially malnutrition and gender relations.

In 2011, Aasma Sheikh, a resident of Rafinagar, became the subject of national media attention. She and her infant son featured centrally in a report that was part of the *Hindustan Times* newspaper's 'Hunger Project' (Bhattacharya 2011). She was the mother of Gulnaz, a severely malnourished child who caught the media's attention. Malnutrition here is common, a product of poor sanitation conditions and crippling poverty. Aasma was prescribed medicines to treat the illnesses that she, Gulnaz, and her other children suffered from and which were exacerbated by malnutrition, but the cost of water meant that she could not afford to buy them. She had to spend Rs.30–40 on water per day. She was faced with the choice: water or medicine. The child later died.

While Mumbai receives an average water supply of 200 litres per capita per day, the city's informal settlements receive an estimated less than ninety litres on average. 'Illegal' neighbourhoods such as Rafinagar receive nothing, at least not officially. Families here earn roughly Rs.100–150 per day, a substantial amount of which goes on water and kerosene. Following the *Hindustan Times* report, the state cabinet minister for women and child development, Varsha Gaikwad, visited Rafinagar, but no change followed and the deaths from a combination of malnutrition, poor sanitation, and low incomes have continued. In fact, local public health NGO Apnalaya stated that the situation had worsened.

In Rafinagar, the open spaces in which children play (for example, the *maidan* or garden area and *kabrastan*, a burial ground) and adults and children spend long hours working, including as rag-pickers in the Deonar garbage ground, are also spaces used for open defecation. Rafinagar comprises an older and more established Part 1 and a newer, poorer, and still expanding Part 2. The geographies of open defecation are organised in ways deemed most proper and safe in the context of prevailing social relations and norms. While young children living in Rafinagar Part 1 use the road outside the settlement, other children and most men of Part 1 walk across the road to the garden or *maidan*, a vast open space located behind one of the private toilet blocks. In Rafinagar Part 2, most young children use the adjacent *maidan* (also known as *kabrastan* since the municipal government had earmarked this land for a graveyard), beyond which rises the Deonar garbage ground, Mumbai's largest garbage disposal site. The youngest of children are often made to sit on newspapers and plastic bags just outside the house because of fears (such as their getting bitten by aggressive stray dogs) associated with letting them defecate further away. For men and women, the Deonar garbage ground with its heaps of garbage provided a particularly suitable topography for gendered separations for open defecation. Men often use open spaces at the lower edges of the ground, especially along the water channel along the ground's western edge, while women walk up onto the garbage ground, finding spaces behind refuse heaps or in the ditches created by amassed garbage to shield themselves from prying eyes.

Women and girls often suffer sexual harassment in their search for privacy. Salma explained:

> Our sons and husbands understand that our mothers and sisters go [to the garbage area to defecate]. But [men] come from outside and harass us.... They [drink] alcohol; they do *charas, ganja, solution....* Many rapes have happened. Some parents don't bring it out in the open to protect their honour; they are scared.[2]

Women go in groups when they can but this is not always practical. Often they also waited until cover of darkness, a wait that exacerbates health problems and can also bring its own vulnerabilities to attack (see Bapat and Agarwal 2003; Truelove 2011).

Although using open spaces such as the *maidan* and *kabrastan* that are visible to more people might reduce the chance of sexual assault, securing some kind of privacy for performing bodily functions is a key concern for women. Most women tried to decrease the possibilities of assault by going to the garbage ground with other women and by going before 10–11 a.m., after which garbage trucks began to ply the ground. However, verbal and visual harassment are not easily avoided. One woman resident explained that if one went alone, someone would 'cover your mouth and carry you off'; this, she added, would not happen if two women went together although men might still pass comments and make obscene gestures. It would not be an exaggeration to say that at times some women took a chance on their safety in their search for privacy and to conform to social norms of modesty. Moreover, going on to the garbage ground to find privacy itself posed risks of being bitten by aggressive stray dogs, falling into deep ditches, and sinking into the garbage especially during the monsoons. For many women, disposing human waste involves clambering over the city's solid waste, the same waste the women are often wading through in order to make a living: here, metabolic processes entail an immersion in a back-and-forth between different forms of waste, hazard, and the generation of value from waste. Violence and risk are central to the ways in which human waste – especially for women and girls – is metabolised through urban space and poverty in Rafinagar. This is true not just of some informal settlements, but for how poor residents experience the metabolic commons as they move around the city.

'If we had to pick one tangible symbol of male privilege in the city,' write Shilpa Phadke, Sameera Khan, and Shilpa Ranade in their 2011 book on Mumbai, *Why Loiter?*, 'the winner hands-down would be the public toilet' (2011: 79). Not only is there a profound imbalance of provisions of toilets for women as compared to men in Mumbai, the size, functionality, and location of public toilets are extremely circumscribed. This is particularly difficult for poorer and usually lower-caste women, who find it harder to make use, for instance, of the toilets of hotels or restaurants, and for whom the lack of toilets is, as Phadke Khan, and Ranade put it, 'a reminder of her unwantedness in the city' (Phadke *et al.* 2011: 80). This unwantedness is particularly striking in Mumbai, the city that has the highest number of working women in the country (Patel 2013), and reflects in part the gendered nature of infrastructure provision in the city and expectations about who uses and should be using public space, as well as cultural notions of pollution and the female body associated in particular with Hindu social orders linking caste and gender. Women's bodies, like toilets themselves, are often linked to contamination, dirt, and pollution, meaning that many people are reluctant even to speak about sanitation: 'Women's bodies are associated with bodily secretions – menstruation, ovulation, lactation – seen as sources of ritual contamination at particular times of the month or year' (Phadke *et al.* 2011: 82).

But there are active struggles in Mumbai aimed at reforming and expanding the metabolic urban commons, and which indeed prefigure a different kind of urban commons in that they question India's unequal metabolic modernity and campaign for a more just set of social relations and urban environments. These movements, disparate in form and number, seek out alternative, more socially just, and environmentally safe ways of metabolising human waste, and often seek to do so with a powerful emphasis on the gendered nature of sanitation in the city. In what follows we examine three: a rights-based imaginary of the urban commons, an entrepreneurial imaginary, and a subaltern imaginary.

Commons 1: the right to urban metabolism

The first imaginary of the commons operates in the realm of rights, data, and citizenship. It maps and counts exclusion. The key referent point in Mumbai here is an umbrella group of civil society organisations in Mumbai titled *Right to Pee*.

In 2013, state women and child development minister Gaikwad was petitioned by Right to Pee. The movement has been campaigning for provisions for women in public toilets in the city. Public toilets are legally supposed to be free in the city, but owing to the almost complete absence of urinals for women, most women have to pay Rs.2 to use them. Gaikwad was petitioned by the Right to Pee movement with a list of 50,000 names, gathered mainly from the city's railway platforms, partly because the municipal corporation (the BMC) had been slow to respond to demands for better provision for women in the city. Gaikwad helped introduce the *Maharashtra Policy for Women* in 2013, which mandates the construction of a women's toilet block every 20 km (12.4 miles) – still, to be sure, a considerable distance between facilities, but a very substantial improvement of provisions for women in the city. This move in turn pressured the BMC to act quicker to build women's toilets in the city, although there has been little substantial construction on the back of the flurry of public statements.

Right to Pee cannot be read as a sanitation movement alone. As Sonia Faleiro writes, 'the unprecedented acknowledgement of a woman's right to a public toilet was seen as a victory not just for the fight for better sanitation, but for the women's movement' (2014). The construction of toilets has always been a local election issue in Mumbai and elsewhere in India, but in recent years the profile of sanitation has grown in local elections, and was in issue in the 2014 national election taken up by BJP candidate and eventual winner Narendra Modi, and is increasingly linked in public debate to gender and caste relations, nutrition, education, labour, and the Indian economy.

The Right to Pee movement, as an amalgam of thirty-six community-based and non-governmental organisations, is an attempt to reflect, represent, and extend the urban metabolic commons through using data and forcing the recognition of different groups into a platform for citizen rights. Supriya Sonar of the Committee of Resource Organisations (CORO), an organisation that helped start the movement in 2011, argued: 'Today, the Right to Pee is everyone's campaign – from women fruit vendors to doctors and educationists, to town planners and gender experts' (cited in Patel 2013). Another activist the movement, Usha Kale, comments: 'Ours is a movement of sewage cleaners and sweepers, flower sellers and fishwives. Just the sort of women who are used to a fight' (Faleiro 2014).

But there is also an important class dimension to these gendered politics, and here class can be mobilised to work not just for but *against* the expansive vision of inclusion and provision that Right to Pee campaigns for. For example, when Right to Pee activists stood on railways platforms soliciting signatures in support of the movement, middle-class women often ignored the petition or '[t]hose who stopped only did so to underscore their privilege. "Who needs public toilets when you have toilets in malls?" ' (Faleiro 2014). The cultural politics of 'pollution' inscribed upon women's bodies that Phadke *et al.* (2011) does not play out in the same way for all women and can become entrenched by class politics.

Commons 2: entrepreneurial civil society

The second imaginary of the urban metabolic commons emerges from an alternative civil society position. While Right to Pee is made up of civil society groups, it is a civil society of a rights-based urban metabolic commons, while the groups we have in mind here espouse more of an entrepreneurial imaginary of the urban commons. This is a form of civil society action that imagines an urban metabolic commons in which the knowledge, capacities, and imaginaries of the poor are placed at the heart of the struggle for better conditions. As with Right to Pee there is, again, a strong role here for data: data about the lives of the poor, about their ability to organise, design, and construct, about their capacity to raise and organise funds, and about their ability to work with different authorities – the municipality, the private sector, international donors, and so on. By placing the urban poor at the heart of the process, this is an entrepreneurial urbanism that chimes with the wider urban script of city entrepreneurialism (MacLeod and Jones 2011; McFarlane 2012), but which does so in a way that places the urban poor at the centre of development rather than the periphery of privately-driven real estate development.

In 2007, a Mumbai toilet block was awarded the prestigious Deutsche Bank Urban Age (DBUA) Award. The toilet block is based in Khotwadi, a well-established informal settlement in west Mumbai. The DBUA award is designed to encourage citizens to take initiatives to improve their cities and runs alongside the Urban Age project, a joint initiative of the London School of Economics and Deutsche Bank's Alfred Herrhausen Society. Describing why the award was given for this toilet block, Deutsche Bank wrote that the project: 'is a striking example of the poor helping themselves, and gives the lie to the stereotypical depiction of slum dwellers as helpless or indolent victims' (LSE News Archive, 2007).

The award is far more than just prestige – US$100,000 was given to the community-based organisation that runs the block, Triratana Prerana Mandal (TPM, 'triratana' means three jewels, and for the activists refers to education, sports, and culture), an organisation that has subsequently used the award to help fund the construction of a large community sports centre along the road from the toilet block. This is an award for citizen entrepreneurialism that refuses to *wait* for the state but instead takes matters – the most fundamental of matters – into its own management. Suketu Mehta, author of the celebrated 2004 book on Mumbai, *Maximum City*, and one of the Urban Age judges, described the toilet project as 'an ingenious as well as indigenous solution that needed very little investment and could be replicated in slum colonies around the world' (Mehta 2011: 155).

The award was given not just because TPM has built a well-maintained, clean block in the neighbourhood, but because the toilet block has become an unlikely focal point for a range of social activities. For example, 200 students from around the local area attend basic computer classes at the block (upstairs from the toilets), paying around Rs.750 for a three-month class. More recently, the block has attained solar hot water, set up a biogas plant, started rainwater harvesting, and ground water through boring – all through new city and state environmental funding schemes. The practice of the sustainable eco-city becomes embodied in a slum toilet block and tied to generating capital through waste – a striking contrast to the pervasive representation of slums-as-waste among not just elites, but more generally in India: 'our aim is 0 per cent garbage', one TPM activist said. 'We are making money [from user charges] and reinvesting it', he went on, in everything from a gymnasium, and computer or dance classes, to a plant nursery behind the toilet, and of course the running of the toilet itself. They have gained international funds for equipment, women's empowerment, and sustainable development.

There are other examples like TPM in Mumbai and elsewhere (e.g. McFarlane 2012; Patel 2015). These experiments in entrepreneurial civil society entail an approach to tackling urban poverty that celebrates the knowledge and potential of the urban poor to build and organise their own individual and collective lives and interests. A more just urban metabolic commons here is not one demanded in the form of rights and provisions by the state, but one in which the local, regional, and central state and other actors (e.g. international agencies like the World Bank or the Gates Foundation) are seen as *partners*, and in which the terms of such a collaboration are geared towards the empowerment of the poor in the development process and, by extension, a more effective delivery of development that enhances the state's skills and potential through partnerships.

And yet, what's clear is that there are moments in which the demand from the *state* – the Right to Pee's demand to basic provisions and for the state to commit to what citizens are entitled to – needs to be pursued, and other moments when an approach to partnership is essential, and here the TPM model of working with and exploiting funding and partnership opportunities with the state and international donors is an important lesson. Any temptation to argue that a rights-based imaginary embodies a more principled approach that will lead to longer term success than an entrepreneurial approach that is too easily co-opted misses the larger point, which is that different moments and spaces require different arts of the political. Sometimes, the sorts of partnerships that an entrepreneurial approach can lead to break down, and a politic of defiance – in the form of a demand, and/or of an insistence that a set of conditions or proposals can no longer stand – is the only way forward.

Commons 3: subaltern struggle

The third imaginary of the urban metabolic commons is a subaltern imaginary. There is, of course not one subaltern imaginary (neither is there one rights-based or entrepreneurial imaginary), and there are links across these different spheres. All I want to suggest with 'subaltern imaginary' is a vast realm of thinking and action that largely exists beyond the plan and activities of civil society organisations, but which is vital to the maintenance of at least some form of the urban metabolic commons, and which is important for civil society groups to operate in relation to.

Occasionally, the cultural politics of 'pollution' can be turned on its head, and in ways that are related to but outside the organisational forms of civil society groups, and which are altogether more fleeting than organised movements like Right to Pee or TPM. For example, in 2010 a public toilet block run by a private company in Rafinagar doubled the pay-per-use price from Rs.1 to Rs.2. A group of residents began to protest the price hike, as Mumtaz related:

> The public created a scene. They went and sat down [to defecate] anywhere, in the *maidan* [open ground], on the road, near the clinic.... So that he [the toilet block caretaker] will also not be able to sit there, he will also get the stink, no?[3]

Mumtaz positions smell, not organisational pressure, as key to this political act. This is a form of protest in which an urban collective temporarily constitutes a political moment that dramatises the limited possibilities for metabolising human waste, especially for women. In this act, women are forced to use their own bodies as political agents in their own neighbourhoods, not to expand the urban metabolic commons but merely to maintain the oppressive status quo. Such battles are flash-in-the-pan, but nonetheless essential to the politics of the urban commons.

The protest in the end was successful, a small victory that points to a wider urban footfall of minor politics whereby residents try to maintain conditions or nudge them in a different direction. These are temporary conflicts that resonate with accounts of lower key contentious politics, where urban public spaces become particularly important for pursuing and registering grievances (for example, Bayat 2010). They are part of a longer repertoire of what Sudipta Kaviraj (1997: 110) calls 'small rebelliousness' around sanitation, where improvised defilement itself becomes a political outlet that depends on the power of smell, irritation, and proximity, a politics that works subversively with the demarcation of the female body as polluting.

Small rebelliousness, emergent in metabolic and social desperation, with threadbare but not insignificant wins, are repeated daily up and down the country. Residents often illustrate their own moral economies of the urban commons through such small acts of resistance. For example, three attempts have been made over the past few years to demolish one of the public blocks in

Rafinagar Part 1 and replace it with a private block. Some residents are concerned about having to pay more for sanitation if a private block is built; some are concerned that they would not have access to a toilet at night (Rafinagar's existing private blocks remain closed from 12–5 a.m.). Residents have successfully come together on two occasions to protest against the demolition attempts and drive away the demolition crew, and managed, if temporarily, to obtain support from the municipal councillor.

It is important to remember too that many of these residents experience multiple forms of sanitation exploitation in that they also work with waste, recycling the city's refuse in the Deonar garbage ground. This is also a live politics in urban India that, much like Right to Pee or TPM, calls into question the metabolic settlement of contemporary urban modernity in India and attempts to forge out a more expansive urban commons. For example, in August 2013, to demand rehabilitation and immediate passage of a pending Bill outlawing 'manual scavenging' – cleaning drains and sewers with your bare hands – and providing provisions for retraining, hundreds of manual scavengers, and many of their family members, from across the country burnt their baskets at Jantar Mantar in Delhi. The protests were aimed at the most oppressive working conditions in the country, where workers are squeezed into narrow urban drains, surrounded by raw sewage and toxic gases.

The protest recalled recent tragedies, some of which had received media attention: for instance, three men killed in Delhi in February 2013 while trying to unblock a drain underneath the Indira Gandhi National Centre for the Arts, or two men killed in Chennai two months later attempting to unblock a forty-foot deep septic tank underneath a hotel. The men, typically working without any safety gear, were killed by asphyxiation from polluting gases in the drains – again, a profound illustration of how the poor suffer from metabolising the wastes of modern urban India. No surprise then, that the protest in Delhi was made up not just of workers, but their families too. While many states in India have banned the practice commonly referred to as 'manual scavenging' – always conducted by Dalits – and insisted sanitation workers are issued with adequate safety equipment, in practice the process continues.

Former Prime Minister Manmohan Singh, speaking in June 2011, called manual scavenging 'one of the darkest blots on [India's] development process', but two years later the Supreme Court expressed serious concern at the inordinate delay in passing the *Prohibition of Employment as Manual Scavengers and their Rehabilitation Bill*, developed in 2012 and aimed at amending and replacing the existing 1993 *Employment of Manual Scavengers and Construction of Dry Latrines (Prohibition) Act*. The new bill promises rehabilitation for manual scavengers in the form of training and education grants, but its slow implementation brought some of India's most marginalised, exploited, and desperate workers onto the streets of the country's capital. At the heart of the protest was a contestation of the implicit idea that the use value of their labour, as a resource in common, is restricted to that of capital rather than of communities and families.

The subaltern imagination of the urban commons is often one of survival, of trying to ensure some measure of provision in everyday life, but like right-based and entrepreneurial civil society imaginaries they too have a longer-term horizon, as the struggle for better conditions for manula scavengers shows. The struggle here is in part one of rights, such as the legislative changes scavengers agitate for or the demand residents make for affordable toilets, but it is more than rights alone. Here, the urban metabolic commons is an often desperate struggle of survival, one that does not speak a language of the state, that places death and suffering at the heart of its politics rather than data, and which therefore isn't captured by rights-based or entrepreneurial imaginations of the urban metabolic commons. Sometimes, it is not enough even to demand rights.

Pushed to the limits, people are forced to use their bodies as political weapons, or to burn baskets and shout that enough is enough. This is not the politics of the entrepreneurial civil society activists, though they recognise the cry. It is not the politics of rights-based citizenship movements, though they connect with the struggle and its hopes. Different moments and spaces call for different arts of the political, and the struggle for the urban metabolic commons cannot do without any of these distinct if related imaginaries. It is not the case that one or other is better, but that one or other is more suited to particular groups and contexts, and then may cease to be later. The lesson here is not to find a place of clinical judgement and proscription, but to locate, nurture, and generate still more experiments.

Another metabolic commons?

The struggle for the commons entails a proliferation of imaginaries of what the commons is and might be, and how it might best be achieved, bound by a shared sense that inequalities wrought by capitalism – in Mumbai, a real estate capitalism accompanied by an often aggressive ethno-religious and territorial exclusion (Hansen 2001; Prakash, 2010) – cannot continue. While these different forms of metabolic politics seek to expand the urban commons to poor neighbourhoods, public places, and precarious forms of work, these politics are not about inclusion or protection alone. In these different forms of metabolic politics there is, to use Edgar Pieterse's (2011) phrase, a set of 'social technologies' for another kind of urban commons, one that imagines a more socially just set of social and environmental relations, especially around gender, class, and caste. They force a series of spaces and groups largely forgotten by India's contemporary metabolic settlement onto the local and national stage: workers cleaning sewers, women attempting to afford local toilets and find somewhere clean and safe for themselves and their children, women seeking to move through the city in the knowledge that there will be local public facilities available, residents and activists who recognise that sanitation opens out into a larger set of social and political issues (work, education, gender equality, dignity, participation in public life, and so on). These metabolic politics position the use value of the commons not in relation to capital but in relation to communities, families, and individuals struggling to survive (Gidwani 2013).

This leaves us with a wider question. What sorts of urban commons do we have in mind when we formulate calls for more *just* cities? What and where are our referent points? Pieterse, writing in relation to urban spaces in Africa that have rarely featured on the contours of mainstream urban theory, states: 'I have no doubt that the street, the slum, the waste dump, the taxi rank, the mosque and church will become the catalysts of an unanticipated African urbanism' (2011: 8). Such spaces and processes can provide catalysts or provocations to potentially different lines of urban theorisation and political formulation of the urban commons. Writing about urban waste pickers in municipal garbage grounds in India, Vinay Gidwani (2013) suggests that theory might usefully do more to connect with the life worlds of waste pickers and their interconnections to spatially distanciated relations of capital, labour, and urbanism. In such spaces there may be sources for new ways of thinking about the urban commons and political change which we might think of, to use a phrase of Gidwani's from that paper, as a 'conjuring of the positive': 'I take this conjuring of the positive from what has been cast aside – marginalized, remaindered, and stigmatized – as the primary intellectual and political task of the postcolonial scholar as archivist of the city' (Gidwani 2013).

Understanding the imaginaries of the urban metabolic commons is one important route through which to better appreciate how urban life takes shape in the present and might be reformulated in the future. While we have concentrated here on informal settlements in Mumbai, and while we agree with Pieterse and Gidwani that sites such as the garbage ground are vital spaces through which to develop a better understanding of the urban metabolic commons and its possibilities, it is also increasingly clear that the struggle for the metabolic commons is not restricted to the global South. These stories connect, in different ways and to different extents of course, with increasingly important parts of life on the margins in Western cities – for instance, in the improvised economies and housing left in the wake of austerity urbanism, or in the calculations that increasingly constitute the everyday lives of British families dependent on food banks, or in the aspirations of social movements like Occupy who see an ever-growing disparity between the elite and the rest, an increasing number of whom are struggling to ensure even the basic staples of daily metabolic life.

What do these three imaginaries of the urban metabolic commons leave us with? Clearly, the challenge ahead for these and countless movements like them is necessarily more than the right to public space alone, although that is a part of the struggle for movements like the Right to Pee. More too than a struggle for pipes and sewers and circulatory infrastructures. It is also a politics of the state's approach to 'illegal' informal settlements, the kinds of metabolic conditions the state chooses to prioritise, and enrols issues as complex as food provision and health services to poor neighbourhoods through to gender, class, and caste politics as they are contingently played out across different parts of the city. It is a politics of recognition and rights, of demonstrating that the skills and knowledge of the poor need to be not simply acknowledged but wedged into urban planning, and of learning from and with subalterns in the struggles they face: affordable toilets, clean water, training possibilities for manual scavengers, a convenient place in which women and girls can use toilets, and so on.

At a minimum, understanding people's everyday experiences of, struggles over and perceptions of sanitation, waste, and water as they live and move through the city is crucial if a pluralised metabolic commons is to succeed in the long term. Inspired by the rights-based, entrepreneurial, and subaltern imaginaries discussed in this chapter, critical urban research can help forge a new urban metabolic commons. This is not about prospecting the urban commons as a singular universalist position, but about promoting an urban commons that is radically differentiated by space (e.g. across informal settlements, public provisions moving through the city), social vectors (gender, caste, class, religion, age, etc.), and occupation (e.g. manual scavengers). These divergences are both inevitable and generative, and reveal to us the value of experimenting with different political arts in distinct contexts and space-times. They demonstrate the need for an ethic of critical care from research that would seek to learn from and nourish diversity rather than attempt to legislate for this or that approach.

There is an important temporal element to struggles for the metabolic commons that we need to better understand in both research and practice. The urgency of the metabolic inequalities in Mumbai demand short-term solutions to end unnecessary death and suffering for no other reason that unclean toilets and dehydration. Subaltern politics often operate in this temporal register of immediacy. Alongside this, there is a need to retain a vision for a longer-term investment in both infrastructure and maintenance work. Ensuring effective maintenance of toilets and drains requires several considerable challenges being addressed: an increased provision of piped water, significantly improved working conditions for municipal sanitation workers, the delivery of context-based low-cost sanitation technologies such as eco-toilets and simplified sanitation where possible (Mara 2012), increased connections where possible of toilets to the sewer network to reduce blockages, and an increase in the number of sanitation workers and of adequate machinery (on this latter point, see De Wit and Berner 2009). In this longer-term register, both rights-based and entrepreneurial struggles have a great deal to teach us in terms of developing strategies and visions for how the commons might be not just redistributed but sustained.

Notes

1 We are grateful to Ash Amin and Phil Howell for their comments on the presentation upon which this chapter is based, and for organising the excellent *Shrinking Commons* conference at Cambridge. Some of the data on Mumbai discussed in the chapter emerges from a collaborative research project with Steve Graham (*Everyday Sanitation*, funded by the UK's Economic and Social Research Council, RES-062–23–1669).
2 *Ganja*, *charas*, and solution refer to different intoxicants.
3 This quote was part of an interview with Renu Desai as part of the *Everyday Sanitation* project conducted by Steve Graham and ourselves.

References

Amin, Ash and Nigel Thrift. *Arts of the Political: New Openings for the Left*. Durham, NC: Duke University Press, 2013.

Bapat, Meera and Indu Agarwal. 'Our Needs, Our Priorities: Women and Men from the Slums in Mumbai and Pune Talk about their Needs for Water and Sanitation'. *Environment and Urbanization* 15, no. 2 (2003): 71–86.

Baviskar, Amita. 'The Politics of the City'. *Seminar* 516 (2002), at: www.india-seminar.com/2002/516/516%20amita%20baviskar.htm (last accessed 5 March 2016).

Bayat, Asef. *Life as Politics: How Ordinary People Change the Middle East*. Amsterdam: Amsterdam University Press, 2010.

Bhattacharya, Pramit. 'A Bad Return on Investment'. *Hindustan Times*, 14 October 2011 (updated 21 March 2013), at: www.hindustantimes.com/india/a-bad-return-on-investment/story-yaWIBJqAvAEpuNNhAQNxAM.html (accessed 18 January 2016).

Desai, Renu, Colin McFarlane, and Stephen Graham. 'The Politics of Open Defecation: Informality, Body and Infrastructure'. *Antipode* 47, no. 1 (2015): 98–120.

De Wit, Joop and Erhard Berner. 'Progressive patronage? Municipalities, NGOs, CBOs, and the Limits to Slum Dwellers' Empowerment'. *Development and Change* 40, no. 5 (2009): 927–47.

Faleiro, Sonia. 'For Some Voters in Mumbai, This Election's All About Toilets'. Quartz, 1, 9 April 2014, at: http://qz.com/196893/for-some-voters-in-mumbai-this-elections-all-about-toilets/ (accessed 25 August 2014).

Gandy, Matthew. 'Rethinking Urban Metabolism: Water, Space and the Modern City'. *City* 8, no. 3 (2004): 363–79.

Gidwani, Vinay. 'Six Theses on Waste, Value, and Commons'. *Social and Cultural Geography* 14, no. 7 (2013): 773–83.

Gidwani, Vinay and Amita Baviskar. 'Urban Commons'. *Economic and Political Weekly* 46, no. 50 (2011): 42–3.

Graham, Stephen, Renu Desai, and Colin McFarlane. 'Water Wars in Mumbai'. *Public Culture* 25, no. 1 (2013): 115–41.

Hansen, Thomas Blom. *Wages of Violence: Naming and Identity in Postcolonial Bombay*. Princeton, NJ: Princeton University Press, 2001.

Jeffrey, Alex, Colin McFarlane, and Alex Vasudevan. 'Rethinking Enclosure: Space, Subjectivity and the Commons'. *Antipode* 44, no. 4 (2012): 1247–67.

Kaviraj, Sudipta. 'Filth and the Public Sphere: Concepts and Practices about Space in Calcutta'. *Public Culture* 10, no. 1 (1997): 83–113.

Loftus, Alex. 'Intervening in the Environment of the Everyday'. *Geoforum* 40, no. 3 (2009): 326–34.

LSE News Archive. 'Deutsche Bank Urban Age Award given to Two City Projects which Transform the Lives of Mumbai's Citizens'. LSE News Archive, 2007, at: www2.lse.ac.uk/newsAndMedia/news/archives/2007/UrbanAge2.aspx (accessed 18 January 2016).

MacLeod, Gordon and Martin Jones. 'Renewing Urban Politics'. *Urban Studies* 48, no. 12 (2011): 2443–72.

Mara, Duncan. 'Sanitation: What's the Real Problem?' *IDS Bulletin* 43, no. 2 (2012): 86–92.

McFarlane, Colin. 'The Entrepreneurial Slum: Civil Society, the Slum, and the Co-Production of Urban Development'. *Urban Studies* 49, no. 13 (2012): 2795–816.

Mehta, Suketu. 'Maximum Cities: Mumbai, New York', in Wilhelm Krull (ed.), *Research and Responsibility: Reflections on Our Common Future*. Leipzig: Europaische Verlagsanstalt, 2011, 149–66.

Mumbai Human Development Report 2009, New Delhi: Oxford University Press, 2010, at: http://mhupa.gov.in/writereaddata/Mumbai%20HDR%20Complete.pdf (accessed 18 January 2016).

MW-YUVA. *Slum Sanitation Project: Final Report*. Mumbai: Municipal Corporation of Brihan-Mumbai, 2001.

Parthasarathy, D. 'Hunters, Gatherers and Foragers in the Metropolis: Commonising the Private and Public in Mumbai'. *Economic and Political Weekly* 46, no. 50 (2011), 54–63.

Patel, Sheela. 'How India's Slum and Pavement Dwellers made Toilets Affordable'. International Institute for Environment and Development (IIED) Blog, 5 April 2015, at: www.iied.org/how-indias-slum-pavement-dwellers-made-sanitation-affordable (accessed 25 August 2015).

Patel, Vibhuti. 'Sanitation and Dignity'. *The Hindu*, 24 September 2013, at: www.the-hindu.com/news/national/other-states/sanitation-and-dignity/article5160791.ece (accessed 25 August 2015).

Phadke, Shilpa, Sameera Khan, and Shilpa Ranade. *Why Loiter? Women and Risk on Mumbai's Streets*. New Delhi: Penguin, 2011.

Pieterse, Edgar. 'Rethinking African Urbanism from the Slum'. *LSE Cities*, November 2011, at: https://lsecities.net/media/objects/articles/rethinking-african-urbanism-from-the-slum/en-gb/ (accessed July 2014).

Prakash, Gyan. *Mumbai Fables*. Princeton, NJ: Princeton University Press, 2010.

Soundararajan, Thenmozi. 'India's Caste Culture is a Rape Culture'. *The Daily Beast*, 6 September 2014, at: www.thedailybeast.com/witw/articles/2014/06/09/india-s-caste-culture-is-a-rape-culture.html (last accessed 5 March 2016).

Swyngedouw, Erik. *Social Power and the Urbanization of Water: Flows of Power*. Oxford: Oxford University Press, 2004.

Truelove, Yaffa. '(Re-)Conceptualizing Water Inequality in Delhi, India through a Feminist Political Ecology Framework'. *Geoforum* 42, no. 2 (2011): 143–52.

10 Inroads into altruism[1]

Marilyn Strathern

> [As an approach to] integration in the society of strangers ... [it is always possible] that a politics of care ... can be decentred from considerations of interpersonal obligation and civic orientation.
>
> (Amin 2012: 16, phrases transposed)

Introduction

Despite the numerous qualifications that particular cases bring, there seems an irrepressible momentum to the type of thinking that opposes self-interest to other-interest. It spills over into contrasts between interests that are short or long term, are mobilised for individual or for social and public benefit, or are self-explanatory traits of the human being against virtues that have to be upheld. Conceptually speaking, the antithesis is self-organising, insofar as each element makes inroads into the other only to have its distinctiveness from the other reasserted. This is no less so than in an intriguing area of UK biomedical policy, intriguing for its indication that what might be rhetorically – or politically – persuasive is less than adequate as an interpretation of the lived state of affairs. That state of affairs involves a commoning of sorts.

Supply and demand

Introducing reflections on what she calls 'the transplant imaginary', the American anthropologist Lesley Sharp speaks of the efforts of medical science:

> to generate [technological] alternatives to organs of human origin such that transplant surgeons might one day bypass altogether the capricious supply of those derived from altruistic strangers, kin, friends, coworkers, and acquaintances who donate parts of themselves to patients dying of organ failure.
>
> (Sharp 2014: 8)

'Might one day bypass altogether': the Hardin of nearly fifty years ago would chuckle at such faith in technical solutions, not at the impossibility of trying but at the impossibility of demand being satisfied. However, the vocabulary of supply and demand is as persistent in UK as in USA attempts to meet patient need. By contrast with the issues raised by Fannin in Chapter 11, scarcity remains the overt economic logic, and with the concept of organ shortage comes something not unlike the delineation of a resource at once limited and of access to all.

The phrase supply and demand refers to the proportions of transplants, organs, and still unmet need.[2] The UK NHSBT (NHS Blood and Transplant service) regularly publishes the numbers of those who have benefited, who have contributed as living or deceased donors, and who are on the transplant waiting list. In 2009–10, when achieved transplants stood at 3,700, some 8,000 potential recipients were waiting for an organ, with 2,000 suspended from the list; in 2013–14 with 4,600 transplants, there were 7,000 people waiting, with 3,100 suspended. Donors, both living and deceased, had increased from roughly 2,000 to 2,450.[3] Gains are counterbalanced by factors that keep up the level of demand: frequently mentioned are the fact that people are living longer and particular health problems are on the rise (notably obesity and diabetes). Equally pertinent, as medical developments introduce more kinds of treatments, new demands are created. This is not the place to weigh up these issues; my interest is in the notion that the flow of organs is blocked by something else in seemingly short supply, although it cannot be computed in numbers.

The availability of body organs does indeed seem capricious. Deceased donations of non-storable body parts depend on the medical conditions under which organs are removed, only a fraction of those on the Organ Donor Register (ODR) dying in circumstances where their organs are viable, and only some of those being suitable for cases in hand. The issue of immunological matching is also relevant to living donations. However, the caprice Sharp put in the mouths of enthusiasts for technological solutions lies elsewhere: it is an attribute of the chancy and uncertain social process that turns strangers, kin, friends, and co-workers into donors. In the eyes of professionals and clients/patients alike, the obvious reason is because donors are persons. If on the one hand lie their state of health and the usability of their body parts, on the other hand lie their state of mind and willingness to donate.

Much has been written, popular and academic, on the concept of body parts. From the professionals' point of view it matters that they are interchangeable, and Tsing's (2012) discussion of 'scalability' is apposite. Her example is the early modern experimentation with growing uniform crops that drove colonial sugar plantations, although she could as well have been talking about the agricultural revolution at home: with the correct conditions, the 'same' crop can be grown anywhere. So too, provided the requirements for the correct medical and physiological conditions are met, one person's body part can be replaced by another's. As an ontological premise, the continuity of the natural world is re-enacted in every transplant.[4] The nature of the body imagined thereby has been the subject of cross-cultural comment: Lock (2002: 226) quotes a Japanese cultural critic contrasting the assumption that organs are replaceable parts with the

Japanese assumption that in every part of a deceased body one may find a fragment of the person's mind and spirit.[5] This does not mean that those who occupy
the former position use body parts without thought for the person, but rather that
the person's mind and spirit are brought to bear on them in very specific ways.[6]
Indeed, a person has a distinct interest, we may call it, in his or her body.

An ethical sensitivity in medical or clinical practice is built by law into transplant procedures themselves, both in UK regulation and in numerous global regulatory frameworks all presuming medical intervention by consent. When it
comes to using bodily material, establishing the consent of the person (colloqui-
ally known as the 'donor') from whose body a part will be taken is a crucial
initial step. For living donors, it is necessary to verify that the circumstances
under which they wish to donate are free of coercion; for deceased donors, to
verify that they had made their wishes known during their lifetime, by signing
the national register (ODR), say, or that a person in a 'qualifying relationship' is
ready to give consent from what he or she knows is likely to have been the
deceased's wishes.[7] What is in short supply, then, are not just organs in an
appropriately healthy state ('well cared for during procurement': Lock 2002: 49)
but organs for which consent has been properly obtained. And here, deceased
rather than living donations present, in the view of some, certain untapped
possibilities.

Time and again diverse advocates, professional and non-professional, including those speaking on behalf of the British Medical Association, have argued
strenuously for procedures that would dramatically transform the question of
numbers: a universal opt-out system. The arguments have made some headway
in one part of the UK, and I return later to recent Welsh legislation. At this point,
it is the protagonists' arguments that are interesting. A universal system would
do away with the need for persons to specifically register consent (that is, to opt-
in). It is taken for granted that the deceased have no personal use for their organs
and the ancillary proposition is that they would be content to have them taken to
save the lives of others. Three things are of note. First, potential donors are being
regarded as part of a population of like-minded people; it is assumed that unless
they explicitly say otherwise everyone shares these views. Second, individual
interest survives death, in a form of the deceased's 'will' to allocate the body to
medical use. Third, above all, the donor's consent is not obliterated but is being
presumed. What is significant is that this consent does not need to be re-
established at the time of death: the organs would have *already* been made available.[8] This would be done through the opt-out system itself, a kind of enabling
infrastructure pre-empting further decision-making. It is almost as though
embodied organs could be conceived of as some kind of national resource in
which an individual person had lifetime usufructory rights that death releases for
dispersal.

Would this shift a sense of the body as, precisely, a resource for exploitation,
as critics argue? If so, the suggestions for an opt-out system envisage it as a
resource that is non-commodifiable; on the contrary, they point to certain aspects
of the whole arena of transplantation that are distinctly commons-like. We may

compare this arena with public service provision in the National Health Service at large, which shares the overarching presumption that distribution of health care is based on patient need. Yet the public goods of the NHS (medication, care procedures, surgical apparatus, all commodities researched, planned, funded) are so evidently finite that they cannot prevent such expectations being constantly shrunk by economic and financial considerations, postcode lotteries, management reforms, and the like. On the other hand, organ transplantation may affect only a few but it emerges as a relatively high-profile arena for restating a commonality of interest in a resource whose scarcity seems equivocal: potentially limitless in nature, scarce for technological and social reasons.[9] In their embodied state, the body's organs are not goods, let alone commodities.[10] Are arguments for opt-out one consequence of a transplant imaginary that sees organs, up until the moment of extraction, as natural entities, material unclaimed, in the 'wild'?

To ask the question obviously leaves the vernacular voice I have until now been trying to convey. However, this is the juncture at which to step to one side of the policy-oriented language in which organ transplantation has been discussed in order to open up a particular theoretical interest in it. What is the social counterpart to such an imaginary – that is, what kind of sociality, what kind of concern or disregard for others, is presumed in such an imagining of organ availability? Scarcity may be the overt preoccupation of those involved in transplantation procedures. However a point long established in anthropological discussions of property, applied by Demian (2004: 65–6; and see Gibson-Graham *et al.*: Chapter 12) to a discussion of the non-excludable nature of common resources, emphasises that it is relations between persons that identify resources in the first place. Here we might turn to something else in seemingly short supply. This is 'altruism'.

Altruism

The issue is the concept of altruism as invoked in the discourse of transplantation, at once among professionals and in the population at large. For present purposes I ignore its currency elsewhere (in philosophy, psychology, and the like) and avoid using it as analytic (I am not concerned with what is or is not altruistic behaviour).

Technical transplant terminology has long applied the idea of altruism to living donations. 'Altruistic' is an epithet for donations to a common pool, the donor thus being in no relation with any potential recipient. A distinction has recently crept in between 'non-directed altruistic donation', where the recipient remains anonymous, and 'directed altruistic donation', where donation is made to a known person. The latter simply underlines what altruism conveys in these circumstances, for the relationship between donor and recipient is not in any meaningful sense prior to the donation itself, which is the only cause of the relation between them. In this usage, then, altruism means that there is no pre-existing relation that is being served by the act. If the consequence of donation is that another's interests are met through sacrifice of one's own, the two sets of interests should not be joined together. The motivations or intentions on each side are irrelevant to the act. In its advice to non-directed altruistic donors to think carefully about their decision, NHSBT states:

> you need to think about the fact that you have no relationship or emotional link with any of the recipients or donors involved in the chain of transplants that may follow your donation. This means that you do not experience the pleasure of seeing a loved one benefit [from your donation].[11]

Otherwise put, this is a mandated case of collaboration between strangers; indeed, the alternative rubric to 'altruistic' is 'stranger' donation.

But how strange is the stranger? Ordinarily the very term donation points to a nexus of concerns where considerations of motivation or intention are intrinsic. As articulated in medical ethics at large, altruism here acquires a whole second set of connotations, applicable whether organs come from living or from deceased bodies. This kind of altrusim is imagined as an internal disposition expressed in people's thoughts or feelings, which will affect their behaviour. At its largest an other-orientation embraces society, as in the celebrated words of one of the first in the UK to donate a kidney to a stranger because 'she wanted to give something back to society' (BMA 2012: 84). Insofar as such a society is imagined in terms of a flat (non-heterogeneous, non-stratified) sharing of interests, commonality is implied. Commonality aside, there is also a specific sense in which the interests of donor and recipient may in fact be imagined as joined, in which the pleasure of seeing a loved one benefit seemingly becomes a template for anonymous interchange as well (e.g. Healy 2006: 116).

The gift-of-life language that saturates transplant talk summons up the very opposite of technical altruism, namely a concrete image of interpersonal relations, 'obligations' even, both colouring the voluntary nature of the act implied in securing consent and personalising the outcome of the benefits. A beneficiary, a recipient, can be visualised. The gift supposedly matches a mental orientation on the part of the donor, who demonstrates that he or she is unselfishly thinking of others as other persons.[12] To a voluntary act is added selflessness, then, people's *willingness* to incur a cost to themselves with others' interests in mind. Altruism, in the view of those who endorse it in this second sense, is not at all to be taken for granted, but should be cherished as a moral disposition, a positive virtue to the extent that an ethical principle is made of it. The very existence of such a principle implies a community of interests between those who share it. Already in one direction donor and recipient are joined in the knowledge each has of the nature of medical need (ultimately, they have the same kind of physical body), although there is a difference of opinion as to whether recipients should be 'deserving' or not, that is, share a like impulse to think of others.[13] Now, in this other direction, insofar as altruism is taken as a personal virtue with public consequences, its enactment may well summon the imagined presence of a likely type of recipient. This is so even when it is given an unambiguous civic cast. Where a general disposition to help others is interpreted as an orientation towards the good of society at large, and thus as bestowing public benefit, particular acts can entail the visualisation of particular categories of beneficiary. Indeed, while such thinking in the abstract encourages impersonal charitable outreach to anonymous recipients, present-day publicists often try to re-personalise that outreach as an imagined donor–recipient relation. Although discouraged by procurement specialists in the early years of deceased donation, professionals today admit the extent to which their patients/clients engage in such personalisation. A visualisation of persons may thus lead to a visualisation of relations; donor kin and recipients (as reported from the USA) adopt all kinds of 'creative strategies for defining their relationships' (Sharp 2006: 171). A joining of interests.

Under this second connotation, altruism may itself be treated as though it were a common good. It has a resource-like aspect, inasmuch as the discourse around organ shortage also entails a sense that altruism might be in short supply. That it *has* to be nurtured or protected is one of reading the Human Tissue Act's (2004)[14] prohibition on commercial dealings in human material for transplantation, the stance also taken in soft law regulation by the 1997 Council of Europe's Oviedo Convention, the 2008 Declaration of Istanbul, the EU Organ Directive, among others. The WHO's guiding principles make explicit reference to 'the societal recognition of the altruistic nature of cell, tissue and organ donation' (NCOB 2011: 69). With no expectation of material return, the gift of an organ must come from altruistic motives. But how could such a generous disposition or other-directedness ever be in short supply? Surely it has no limits?

The answer is already anticipated in these prohibitions. Because of the oft-voiced antithesis between self-interest and other-interest, altruism is regarded as an impulse (to behave in a particular way) in competition with diverse other impulses that flourish to quite opposite ends, those that feed selfishness. In this nexus of concerns around organ donation, self-interest carries a negative value, one invariably expressed through the concrete image of commerce. Money stalks the very definition of the gift. Public consultations on donation practices meet vexing and unresolved questions about the cost of making a donation (paying expenses), or the kind of debt that recipients of cadaver organs feel (a need to make some kind of return). Such transactions may be interpreted as the thin end of a sinister wedge that would erode the virtually ubiquitous international prohibition on organ purchase. Gift-giving lives under the constant shadow of a market alternative, so much so that the slightest suggestion of monetary recompense for donors can lead to expressions of moral outrage. Here the virtues of other-interest may be closely allied with disinterest. In the discursive arena of transplant practices, altruism (a personally disinterested interest in others) and money (a means of self-interest) are in open conflict.

By no means the only way of assembling the various elements involved in organ donation, this antithesis dominates as a model of its transactional aspects.[15] The background assumption is that money invariably has the trumping hand, while altruism invariably needs propping up. Self-interest requires no champions;

it is often invoked as a default position of 'being human'. At the same time, altruism is prone to perpetual shrinkage. The evidence lies in the very extent to which people have to cherish it, by voicing its value, encasing it in regulation, in short, working on its behalf. Observing such exertions, someone from outside the model might say that the more altruism is explicitly talked about, the more it is implicitly presented as under threat. It is in this respect, it seems to me, that altruism is treated as though it were a non-excludable resource but for the exclusions created by contrary impulses. Unlike organs, it could not be said to exist in the wild: it has to be intensively cultivated, flourishing only if everyone contributes to it. The capacity for altruism, in this view, might be exercised for the common good in that any person might benefit from other people's altruistic impulses, but these expectations are liable to disappear unless safeguarded. Inroads come not from those who use the resource in such a way that the cost is concealed in being borne by numerous others; they come rather from people's failure to articulate its value or live up to its moral imperative and *thus make it into* something all have in common.

Inroads also come from that other quarter, from the language of money, assumed these days to be ever-expanding, from the financialisation of all kinds of decision-making to the ability to put a figure on anything. An explicit argument heard over and again against money or other forms of material recompense (which can always be given a monetary value) being used to increase the supply of organs is that anything like remuneration would undermine or crowd out the altruistic impulse.[16] In a model where money is an index of self-interest, remuneration would turn a selfless commitment to society at large into a routine reward for services. Hence, some fear, were legislation to make it possible for someone or their relatives to be remunerated for donating, each remuneration would in effect decrease ('crowd out', 'weaken') a principle based on other-interest; the idea of being alike in basic need, 'everyone in it together' in an oft-quoted phrase, would be eroded. Other-interest is axiomatically most evident when it is held in common. This contributes to the sense in which altruism is itself a common resource on which anyone might one day find themselves depending.

It is intriguing, then, that exactly these criticisms about crowding have been made in relation to schemes for opt-out, that is, where consent is presumed unless the contrary is stipulated. In 2008 the Department of Health's Organ Donation Taskforce was asked if it would recommend an opt-out system for the use of deceased donor organs across the UK. It rejected the idea, concluding that such a system would undermine the concept of donation as a gift, erode trust in practitioners, and actually reduce rather increase donor numbers (NCOB 2011: 103). Now, with legislation (the Human Transplantation [Wales] Act 2013) that came into effect in 2015, the Welsh Assembly is in the process of implementing an opt-out scheme for Welsh hospitals. It formulates a 'soft' procedure, in which the presumption of consent ('deemed consent') is backed by consultation with the deceased's families. This is to confirm what is known about the donor-to-be's wishes.[17] Consent is thus retrieved as a paramount consideration. At a minimum it is arguable that information about such wishes helps conserve the justification of using the word donation: organs are deemed given, not taken.

The Organ Donation Taskforce was earlier fearful of an encroachment on altruism. Yet what we have here seems an interesting limitation to the very model of an antithesis between self- and other-interest. Insofar as everyone is given the opportunity to opt-out of automatic donation, everyone is also given the opportunity to stay in. There is no need to appeal to a heightened sense of other-interest as an incentive (and money drops outs of the picture).[18] Individual and thus heroic acts of altruism are no longer visible, and by the same token we might question whether the opting-out population is embodying self-interest (in any strong sense) any more than the UK majority who decline to join or ignore the Organ Donor Register. Indeed, opt-out suggests itself as an example of what, in quite other situations, Amin's *Land of Strangers* (2012: 50) envisions as the sociality underpinning a politics of care, where he talks of strangers becoming collaborators. Perhaps operationalising presumed consent, as the opt-out logic has it, simply takes to an extreme a practical or organisational possibility for shaping care for others, a collective labour of sorts, of the kind that surfaces from time to time in organ donation procedures generally. The role being ascribed to families, however, introduces a further turn in the topic, and an altogether different limitation to the antithesis.

Enter kinship

In what sense is altruism a limited resource? It may appear so in the (antithesis-driven) model of values-erosion described here, and more generally perhaps in debates based on the premise that ethical values must be made visible or else they will (of course) disappear. Altruism seems always in danger of disappearing. Yet if we look at people's actions over the last few years in the UK there has been a steady if small rise both in those joining the ODA and in those who, whether as living or deceased donors, have contributed to a rising number of transplant operations.[19] Now something approaching a half of all donors are living (donating mainly kidneys, but also livers and, rarely, lungs). While a tiny but annually increasing number come forward for 'non-directed' transplants, that is, to strangers, the majority involve donations to people who are known to the donor, in short, relatives or friends. No shrinking of a commons, no shortage of altruism here! Nonetheless, the involvement of kin relations gives one pause.

An observer might argue that when it comes to relatives, it is actually non-sense to disentangle self-interest from other-interest. Between kinspersons, reciprocities that might look for all the world like tit-for-tat transactions are intrinsic to the ongoing nature of relationships, the heterogeneous temporalities of give and take. Insofar as thinking of others and thinking of oneself run together, organ donations among relatives confound the other-interest versus self-interest model that dominates so much of the discussion. Kinship, the observer might remark in parenthesis, also cuts across any simple division of public and private interests. And as to what is 'self' or 'other', if the observer is an anthropologist he or she might quote Aristotle's dictum of kinship as 'the same entity in discrete subjects' (Sahlins 2013: 20), or in defining personhood find in kinship 'the relations that are perceived as being intrinsic to the person' (Carsten 2004: 107).[20] Such mutuality may be to expansive or to coercive effect. Kaufman voices the reasoning of a reluctant organ recipient: 'my family needs and wants me to live.... Therefore I need to live, so they ... will offer to donate a kidney for me, and (though it may not seem right) I must accept it' (2009: 33).

Money ceases to have quite its demonic signature. Ties of kinship are by and large not appropriately expressed in market terms, yet finance plays a huge role (Zelizer 2005; McKinnon and Cannell 2013). One only has to think of household provisioning, inheritance, present-giving. Spending on others (who are close to one) is in a sense spending on oneself (who belongs to others).[21] To the extent that a relative is already part of the donor, then donation between relatives is for the donor as well as for the recipient, as the awkward reluctance of some recipients, and shared joy of others, attests. So if, in the context of living organ donation, payment within families is inappropriate, it is hardly because altruism needs to be protected. Rather, the contrast between altruistic reasons and selfish reasons is thoroughly ambiguated. I do not mean to imply that individuals cannot act in all kinds of positive and negative ways towards their relatives, including with lesser and greater degrees of selfishness, but when they act *as relatives* their actions raise very specific expectations. Doing everything for one's child or nothing for it translates neither into simple other-interest nor into self-interest. It is thus uncertain where the self-interest of a selfish mother (that is, in relation to her children) is to be located. Of course, to continue with this example, not all self-interest is in the ordinary sense 'selfish'. Is a concept nearer to enlightened self-interest the answer? This would imply a fusion of selflessness and selfishness, echoing the particular kind of market logic that sees private benefit as also being for public benefit, first voiced – with enduring consequences – at the time of the Scottish Enlightenment (Adam Smith's 'invisible hand' by which self-interest promotes others' interests). For Hardin (1968), of course, assuming that the best decision for the individual, locked as he or she is into a specific situation, will be best for all was the fatal flaw in thinking about limited resources. We return to Enlightenment arguments, but for the moment note that kinship affairs are far from the operations of an invisible hand. On the contrary, the relations that families and relatives maintain are highly personalised, one guarantee of continuing relations among kin being that reciprocities over time are open-ended and visible. Much that is bundled up in the self-interest/other-interest antithesis simply seems not to apply.

The same is also true of another category of interpersonal relations: friendship. And here too relations are visible. As with kin, and by contrast with acquaintances, friends not only enjoy one another's goodwill but crucially know of the basis for their interaction and each 'knows that the other knows it' (Beer and Gardner 2015). Indeed, in some respects there appears no difference between relatives and friends, as the facilitators of organ transplants acknowledge. The NHSBT website describes living donors as likely to be close relatives or 'individuals who are not related but may have an established emotional relationship with the recipient such as a partner or close friend' (NHSBT 2015). The list of persons able to give consent in organ donation because of their relationship with the deceased similarly pairs relatives and friends, a bracketing endorsed by the wording of the Welsh legislation. Thus instead of consent being deemed, someone may prefer to voice 'express consent' by appointing a 'relative or friend of long standing' as a representative who will know the deceased's view. Concomitantly, a 'relative or friend' with such knowledge may object to consent – not just anyone who knows the wishes of the deceased, but if not a relative then specifically 'a friend' whose relationship implies long-term familiarity.[22]

Now the antithesis between self-interest and other-interest has been the focus of an argument from sociology about the very category of friend in the political and philosophical work of early modern writers, insofar as they contrasted a new commercial society with pre-commercial societies that could offer no possibility of disinterested relations. Speaking of scholars of the Scottish Enlightenment, Silver posits that their 'theoretical project requires an enlarged domain of indifferent persons available for market exchanges and contractual engagements' (1990: 1491). It was 'authentically indifferent co-citizens' who, in this imagining, peopled the then new 'strangership' of commercial society (Silver 1990: 1482–3). Encouraging the exercise of anonymous self-interest, this (commercial) sphere was thus identified with the development of instrumental relations between notional strangers, explicitly removed from the restrictions of pre-existing bonds – and from the mutual interests that strengthened them – once held to engage kin and friends alike. At the same time, he famously argues that these writers imagined commerce as precipitating a second, distinct, sphere of 'personal and civic friendship', 'a new concept of non-instrumental personal relations' based on common sympathies (Silver 1990: 1486, 1492). Applied alike to intimate sociability and to generalised benevolence towards others, it fostered the development of friendship for its own sake.[23] In the early formation of this antithesis to self-interest, the language was as much of disinterest as of other-interest.

The anthropologist might add that this Enlightenment vision required the conceptual detachment of persons, as individual selves, from relations. Thus the interests of established relationships had no part to play in the indifferent ('disinterested') marketplace precisely because 'self-interest' did, while conversely the individual friend was selfless ('disinterested') only to the extent of downplaying self-interest, for he or she would be investing in the interests of the very relationship itself.[24] In hindsight, arguably, some of the ambiguities of self-interest and dis- or other-interest turned on whether interest was attached to the person or the relation.

Kinds of commoning

This brings the narrative to an interesting conjunction.[25] One might argue that an emphasis on altruism belongs precisely to the sphere of rhetorical solidarities upheld in the image of interpersonal obligation and civic orientation that Amin sees as a distraction to contemplating already-present and future of modes of sociality. The alternative is to identify the functionality of social formations through their infrastructures, through collective labour in the contingencies of situated practices. Organ transplantation discourse affords an example of rhetorical solidarity, and of the problem it creates. If altruism's constant battle against self-interestedness and commerce is a sort of experiment in strangerhood, it is one that constantly trips itself up, precisely because the battle is all-consuming. However, the same discourse has also led to developments along the lines of new functionalities: the Welsh opt-out scheme as an infrastructure of (not quite automatic) enablement. The scheme makes more clear not less where common interests lie – and it side-steps the antithesis altogether. In fact, I hazard, there might be kinds of commoning defined through their very suspension of that antithesis.

This chapter has toyed with the idea of (embodied) organs and (impersoned) altruism being commons-like resources, insofar as they are at once available to anyone and, although treated as always in short supply, not otherwise discrete as goods or commodities. However, this is not the burden of my conclusion. Rather it points to the implications of the brief foray into another arena where the self-/other-interest antithesis hardly applies.

The technical phrase 'altruistic' or 'stranger' donation for living organ donation to a common pool denotes a kind of collaboration between strangers, an impersonal one by contrast with collaborations between those who, already known to one another, enact out their personal relations at the same time. The logic of the situation allows the observer to reflect back on what is happening when the latter are donors and recipients, as kin and friends may become. Insofar as the paradigmatic organ transplant has been (deceased) transfer between strangers, as blood donation was originally conceived between strangers (Whitfield 2013), perhaps we can also imagine them (friends, kin) acting as 'strangers' would act. This would be to see them as though they were a kind of non-anonymous counterpart to today's 'intimate publics', which Amin (2012: 30–1) discerns in the society of strangers. For we are not obliged to think of friends and kin only in terms of the compulsions of community, of the solidarities and group allegiances that get in the way of imagining new forms of social living. But there is a question about where to put relationships.

I deliberately did not refer to the enabling infrastructures of opt-out organ donation as 'impersonal', for the personal–impersonal axis simply reinvents the old antithesis.[26] A severe problem generated by these antitheses is the imagination of social relations, the kind of sociality they project.[27] Here is a piece of Enlightenment legacy that English-speakers might wish to reconsider. It concerns not the well-known detachment of person (as conscious self) from body, and we have seen how today that is taken care of in transplant protocols, but the detachment of person (as individual) from relations. Such detachment was openly promoted to the invisible benefit of all, so that 'relations' – family, friends – must flourish only for themselves; these days, the technical usage of 'altruism' explicitly brackets off pre-existing relations. As organ donation has developed into its present lived state of affairs, however, perhaps we have chanced on ways of acting, and a capacity for action, that require perceptions of relationships – not just of the 'other' person – that take kinship and friendship in new directions.[28] Intimate strangers, collaborators in one another's projects, living one another's lives?

We can imagine, with Amin, present-day epistemic communities coordinated by interactive technologies, where 'coalitions are short-lived, the individuals self-centred, the work divided, the identities formed elsewhere [where] ... strangers become collaborators but not friends, co-generators of often quite extraordinary innovations, but without interpersonal ties' (Amin 2012: 49–50). What this chapter might have interpolated is a question about where to put those excluded capacities for friendship and interpersonal ties, extending as they do over time. Kinship and friendship do not of themselves need to entertain appeals to common humanity or to the politics of community solidarities. At least, this is arguably the case insofar as kin and friends alike are indifferent to the antithesis of self-interest and other-interest. Rather, an infrastructure of enablement, to good or ill, and whether or not to anyone else's benefit, lies in their very relation to one another. They do not have to find any other reason to act. Such a 'coming together as one', as the editors phrase it, has always been the case. If what is held in common is produced through the relationship as such, conceivably we have chanced on a(nother) resource for thinking about future commons more generally.

Notes

1 My warm thanks to Hugh Whittall and Katherine Wright of the Nuffield Council on Bioethics for encouraging me to write on this topic. In 2010–11, I chaired the Council's Working Party on organ and tissue donation (NCOB 2011). The chapter draws on no information that is not in the public domain; the conclusions are entirely mine and not attributable in any way to the Council or its directorate.
2 Many transplant professionals see this as the most succinct way to talk about waiting lists, but it is generally widespread in public discussions about health funding allocations.
3 There is no match between individual donors and transplants: in deceased donations more than one organ may be taken from a body. Over the same period the number people signed up to the ODA increased from eighteen million (about 30% of the UK population over eighteen) to 20.2 million; by April 2014 it was twenty-one million. (The NHSBT was actively campaigning over this period. This information comes from NHSBT annual reports as summarised on the NHSBT website. [2009–10 figures are reproduced in NCOB 2011: 87; 2013–14 activity report, accessed 27 March 2015]).
4 The allusion here is to 'naturalism', in Descola's (2013) fourfold scheme of world cosmologies, or to Viveiros de Castro's (2014) mononaturalism as opposed to the multinaturalism of Amerindian perspectivist thinking.
5 Drawn from a work of Yonemoto Shōhei published in 1985. Lock is well known for demonstrating the local nature of physiologies created by specific conditions, as opposed to notions that ascribe universality to the body's workings. This is akin to Tsing's advocacy of the 'non-scaleable', indicating entities with relations attached. To opposite effect, planters were experimenting with types of cane and soil in order to facilitate the interchangeability of forms; the varieties so propagated were genetic isolates without interspecies ties. 'One must create *terra nullius*, nature without entangling claims' (Tsing 2012: 513, emphasis in original).

6 The contrast is specifically with Americans but would include the English too. On French exceptionalism when it comes to notions of body and person, see, for example, Dickenson 2007.

7 The UK Human Tissue Act 2004 sets out a list of such qualifying relationships, first relatives in terms of distance from the deceased, then 'friend of long standing' (see NCOB 2011: 59).

8 When time is at a premium. As we shall see, the Welsh scheme is a 'soft' version, which does re-establish the issue of consent insofar as families are routinely consulted.

9 This is separate from the fact that organs fit for transplant only exist through immense investment in (the research, productivity, financial activity of) transplant services (Waldby and Mitchell 2006).

10 And in the UK legally speaking the property of no one, including the person to whom the body 'belongs', despite widespread notions of ownership to the contrary. It is in the context of ideas about property that questions explicitly address the language of the commons; see for example Dickenson's (2007) discussion, which takes into account diverse arguments over applying concepts such as *terra nullius* or the 'new enclosures' to bodily materials. However, property is a dimension that must be left to one side here. With reference to body parts, it is taken up in Chapter 11 by Fannin; both she and Gibson-Graham, Cameron, and Healy (Chapter 12) would challenge the scope of the scarcity model that is my ethnographic starting point insofar as it underpins current approaches to organ procurement.

11 NHSBT website 'Organ donation: non-directed altruistic donation', at: www.organ-donation.nhs.uk/how_to_become_a_donor/living_donation/national_living_donor_kidney_sharing_scheme/non-directed_altruistic_donation/ (accessed 27 March 2015).

12 The gift is here understood with its vernacular connotations of something freely given and with no expectation of return (the last being contentious, but ideally summoned when the obligation to return is taken as extrinsic rather than intrinsic to the gifting). The gift is thus detached from the person; elsewhere (e.g. Strathern 2012) I have pointed to the severance work that this kind of detachment does in rendering organs ethically fit for transplanting, the medical need for the differentiation of the organ from the individual person who supplies it being discussed by, among others, Lock 2002, Waldby and Mitchell 2006, Healy 2006.

13 They are joined *by* that medical knowledge. Apropos 'deserving', people may equivocate about the ethics of donating to those who have brought their medical condition on themselves through indulgence/abuse or 'selfishness'.

14 Applicable to England, Wales, and Northern Ireland, Scotland being covered by the Human Tissue (Scotland) Act 2006.

15 There is a wide range of positions voiced in the UK on the issues described here, and many potential models. But whether they subscribe to it (aligning the actions of themselves or others by it), or vehemently deny it (the antithesis need not or does not exist), at present the antithesis itself seems a foundational axis to many people's mobilisation of the diverse values they bring to the subject.

16 Here, and in other references to public or professional opinion, I draw on the information and contributing literatures summarised in NCOB 2011. (The remit of that report included an enquiry into the role of material and immaterial incentives in promoting bodily donation). It would be an interesting but separate enterprise to triage these discussions in the public domain, so-called, with respect to the diverse circumstances of the UK's many populations.

17 The procedure adapts the already existing convention of qualifying relationships (see note 7) in order to ascertain if there is any objection to consent being deemed in place.

18 The antithesis is equally rendered irrelevant at the opposite extreme, the all-money market model for increasing organ donation (e.g. Harris 2003; Erin and Harris 2003).

19 Much of this can be put down to the concerted campaign by NHSBT (see note 2). Living donor numbers in the UK increased from 858 (2007–08) to 1,045 in 2010–11, to 1,136 in 2013–14 (in 2013–14 deceased donors numbered 1,320). The increase in numbers of living donations is facilitated by medical developments reducing the trauma of the operation for the donor.

20 It is no surprise that the antithesis is elaborated in contexts where the individual (as a psychological or legal entity, say) is a significant descriptive of the person. Sahlins continues: 'kinsmen are persons who belong to one another, who are parts of one another... whose lives are joined and interdependent' (2013: 21). Needless to say the realities play out in myriad ways, and the extent to which the ideas sketched here inform specific situations must be, as Edwards (2009: 5) would remark, ethnographic questions.

21 'Close' should be divested of its affective connotations, for this will otherwise be read as a sentimental view of kin relations. Disruption and avoidance between kin may be as much a mark of 'close' kinship as the opposite. In this cosmology, kin become disengaged from one another as kin to the extent that they bring to the fore other aspects of their social being, for example, the capacity to act as independent individual persons.

22 The potential proximity of kin and friends (family as 'friends', friends as 'family') is relevant. Kaufman's (2009) study of living donation in the USA reports the sentiment of a woman who donated a kidney to a friend: there was no question because he was 'like family'.

23 'Only with impersonal markets in products and services does a parallel system of personal relations emerge' (Silver 1990: 1494). He observes that, by contrast with later developments, these Enlightenment thinkers saw 'commercial society' as positively releasing personal relations from their former instrumentalism; antithetical in character, the two spheres nonetheless supported each other. Those later developments gave social science, among other things, both an enduring, albeit contested, analytical contrast between 'the public domain of economics and politics' and 'the private domain of kinship', as Bodenhorn (2013: 131) summarises it, and 'the personal and non-institutional character of kindness' as a philosophical entry into the concept of cosmopolitanism (Josephides 2014: 1).

24 Silver's (1989: 277) own exegesis (albeit of modern friendship) ignores the relationship as an object of interest.

25 In the theoretical understanding of the sociality at stake, not in any policy-informed sense. The tension is to be lived with, for this is not the time to jettison the public discourse of 'altruism', any more than that of 'human rights', or indeed any more than feminist activism could ever dispense with the often analytically dubious concept of 'women'.

26 The personal/impersonal antithesis fuels the conventional late twentieth century/early twenty-first-century audit-driven 'distrust' of people's commitment to institutional projects, and the motivation to put 'impersonal' measures in their place. The issue is simply noted; in the present political climate, it would be as hazardous to do away with the 'impersonal' as a public virtue as it would be to do away with 'altruism' (see note 25).

27 I use this old fashioned phrase, as throughout this chapter, to suggest an emphasis on relations between persons, since of course relationality as such is a pervasive and ubiquitous property of systems of all kinds.

28 Whether to lament or celebrate, a radical example of innovation in relations lies to hand. Over the English-speaking seventeenth and eighteenth centuries, 'friend' changed from connoting a closely related supporter, including the most loyal of one's kin, to being the archetype of the disinterested (non-kin) relation perpetuated for own sake. (There were concomitant changes in conceptions of kinsfolk).

References

Amin, Ash. *Land of Strangers*. Cambridge: Polity Press, 2012.

Beer, Bettina and Don Gardner. 'Friendship, Anthropology of', in James D. Wright (ed.), *International Encyclopedia of the Social & Behavioral Sciences, Vol. 9*, 2nd edition. Oxford: Elsevier, 2015, 425–31.

BMA [British Medical Association]. *Medical Ethics Today: The BMA's handbook of Ethics and Law*. 3rd edition. Chichester: Wiley-Blackwell for BMA Medical Ethics Department, 2012.

Bodenhorn, Barbara. 'On the Road Again: Movement, Marriage, Mestizaje, and the Race of Kinship', Susan McKinnon and Fenella Cannell (eds), in *Vital Relations: Modernity and the Persistent Life of Kinship*. Santa Fe, NM: School for Advanced Research Press, 2013, 131–54.

Carsten, Janet. *After Kinship*. Cambridge: Cambridge University Press, 2004.

Demian, Melissa. 'Seeing, Knowing, Owning: Property Claims as Revelatory Acts', in Eric Hirsch and Marilyn Strathern (eds), *Transactions and Creations: Property Debates and the Stimulus of Melanesia*. Oxford: Berghahn, 2004, 60–82.

Descola, Philippe. *Beyond Nature and Culture*. Translated by Janet Lloyd. Chicago, IL: University of Chicago Press, 2013 [2005].

Dickenson, Donna. *Property in the Body: Feminist Perspectives*. Cambridge: Cambridge University Press, 2007.

Edwards, Jeanette. 'Introduction: The Matter in Kinship', in Jeanette Edwards and Carles Salazar (eds), *European Kinship in the Age of Biotechnology*. Oxford: Berghahn, 2009, 1–18.

Erin, Charles and John Harris. 'An Ethical Market in Human Organs'. *Journal of Medical Ethics* 29, no. 3 (2003): 137–38.

Hardin, Garrett. 'The Tragedy of the Commons'. *Science* 162, no. 3859 (1968): 1243–8.

Harris, John. 'Organ Procurement: Dead Interests, Living Needs'. *Journal of Medical Ethics* 29, no. 3 (2003): 130–4.

Healy, Kieran. *Last Best Gifts: Altruism and the Market for Human Blood and Organs*. Chicago, IL: University of Chicago Press, 2006.

Josephides, Lisette. 'Introduction: We the Cosmopolitans: Framing the Debate', in Lisette Josephides and Alexandra Hall (eds), *We the Cosmopolitans: Moral and Existential Conditions of Being Human*. Oxford: Berghahn, 2014, 1–28.

Kaufman, Sharon, Ann J. Russ, and Janet K. Shimm. 'Aged Bodies and Kinship Matters: The Ethical Field of Kidney Transplant', in Helen Lambert and Maryon McDonald (eds), *Social Bodies*. New York: Berghahn, 2009, 17–46.

Lock, Margaret. *Twice Dead: Organ Transplants and the Reinvention of Death*. Berkeley, CA: University of California Press, 2002.

McKinnon, Susan and Fenella Cannell (eds). *Vital Relations: Modernity and the Persistent Life of Kinship*. Santa Fe, NM: School for Advanced Research Press, 2013.

NHSBT 'Organ donation: non-directed altruistic donation', 2015, at: www.organdonation.nhs.uk/how_to_become_a_donor/living_donation/national_living_donor_kidney_sharing_scheme/non-directed_altruistic_donation/ (accessed 27 March 2015).

Nuffield Council on Bioethics [NCOB]. *Human Bodies: Donation for Medicine and Research*. London: NCOB, 2011.

Sahlins, Marshall. *What Kinship Is – and Is Not*. Chicago, IL: University of Chicago Press, 2013.

Sharp, Lesley. *Strange Harvest: Organ Transplants, Denatured Bodies and the Transformed Self*. Berkeley, CA: University of California Press, 2006.

Sharp, Lesley. *The Transplant Imaginary: Mechanical Hearts, Animal Parts, and Moral Thinking in Highly Experimental Science.* Berkeley, CA: University of California Press, 2014.

Silver, Allan. 'Friendship and Trust as Moral Ideals: An Historical Approach'. *European Journal of Sociology* 30, no. 2 (1989): 274–97.

Silver, Allan. 'Friendship in Commercial Society: Eighteenth-Century Social Theory and Modern Sociology'. *American Journal of Sociology* 95, no. 6 (1990): 1474–504.

Strathern, Marilyn. 'Gifts Money Cannot Buy'. *Social Anthropology* 20, no. 4 (2012): 397–410.

Tsing, Anna. 'On Nonscalability: The Living World is Not Amenable to Precision-Nested Scales'. *Common Knowledge* 18, no. 3 (2012): 505–24.

Viveiros de Castro, Eduardo. *Cannibal Metaphysics.* Edited and translated by Peter Skafish. Minneapolis, MN: Univocal, 2014.

Waldby, Catherine and Robert Mitchell. *Tissue Economies: Blood, Organs, and Cell Lines in Late Capitalism.* Durham, NC: Duke University Press, 2006.

Whitfield, Nicholas. 'Who is My Stranger? Origins of the Gift in Wartime London, 1939–45'. *Journal of the Royal Anthropological Institute* 19, no. S1 (2013): S95–S117.

Zelizer, Viviana. *The Purchase of Intimacy.* Princeton, NJ: Princeton University Press, 2005.

11 Revisiting a bodily commons
Enclosures and openings in the bioeconomy

Maria Fannin

The mapping of the human genome is often imagined as the paradigmatic moment of both enclosure and opening in the twentieth-century life sciences: the enclosing of knowledge about the building blocks of 'life itself' through networks of scientific authority, the entrepreneurial aspirations of individual scientists, venture capital, and formal cultures of competition; and the opening of new horizons of human becoming, spaces of hope and anticipation, and global collaboration. Today, bodies, body parts, organs, tissues – virtually any biological material obtained from living bodies – increasingly generate scientific research use value and potential exchange value in what has been called a 'second enclosure movement' (Boyle 2003), rivalling in scale and scope the early modern enclosures of common land into private property. In this context, generating new 'resources' for scientific research relies heavily on legal regimes that protect forms of intellectual property but leave unrecognised and unremunerated certain forms of bodily labour.

This chapter draws attention to commoning, enclosure, community, obligation, gift, and debt in the domain of the life sciences and more fundamentally in the biopolitical spaces and practices that make living bodies, human and non-human, subjects of new forms of biological knowledge and practice. It maps the geographies of enclosures and openings in a 'bodily commons' in two ways. First, through an exploration of how the language of the bodily commons is posed as a critical articulation of what's lost in moves to enclose, privatise, and restrict access to the immaterial and material vitality and generativity of living bodies. The term 'bodily commons' is inspired by Donna Haraway's claim that: 'the gene and gene maps are ways of enclosing the commons of the body – of corporealizing – in specific ways, which, among other things, often write commodity fetishism into the program of biology' (2005: 120). This notion of the commons of the body demonstrates what Michael Hardt refers to as the immanent process of commoning, where a commons is continuously generated by processes of capitalist enclosure. Unlike other spaces and sites at which the enclosure of a commons are frequently analysed (the global, the urban, the domestic, or common-pool resources such as land, water, or other forms of 'common property'), the commons of the body is both *made*, or in Haraway's terms, 'corporealised', and *unmade* or enclosed by the contemporary political economy of the life sciences. In this way, the commons of the body does not represent a commons once shared and now lost, but a future-oriented 'political horizon' (Sevilla-Buitrago 2015).

Haraway's reference to a bodily commons, in the wake of the mapping of the human genome, evokes the immaterial store of the human body's living vitality, translated into 'information' or 'genetic code'. At the same time, the notion of a 'bodily commons' references the very fleshiness of the body, the body as living and breathing, as the *habitus* living beings share: the bodies we are, as well as the bodies we do. Haraway warns against the dehistoricisation or romanticisation of this bodily commons. At the same time, the immanent production of the commons of the body offers alternative, affirmative ways to imagine anew the subjects and spaces of the commons.

Second, the invocation of a bodily commons here points to what is increasingly recognised in the register of global environmental crisis, as the call to conceptualise a shared form of life, what Dipesh Chakrabarty (2009) calls 'species-being'. Reflections on the enclosure or creation of a 'genetic commons' in the immediate aftermath of the mapping of the human genome prefigure this conception of species-being by naming the project as one for all humanity thus '[bringing] into view certain other conditions for the existence of life in the human form that have no intrinsic connection to the logics of capitalist, nationalist, or socialist identities' (Chakrabarty 2009: 217). Species-being invokes the sense of being 'connected rather to the history of life on this planet, the way different life-forms connect to one another' (Chakrabarty 2009: 217). If the concept of species-being speaks to the common future horizon of a shared developmental or co-evolutionary past with other species and other beings, the identity brought into view by the mapping of the human genome is *also* simultaneously a 'biological-being', intimately imbricated with the logics of enclosure that characterise capitalism, nationalism, and socialism, even as it exceeds those logics. The discussion of the enclosure of a 'genetic commons' thus suggests that moving from this 'biological-being' to another emerging conception of a bodily commons requires new vocabularies and imaginaries, new ways of thinking the connections between species and between forms of life.

Much of the work analysing late twentieth-century developments in the life sciences highlights their increasingly marketised and commercial character. As Hilary and Steven Rose write in their trenchant critique of the cultures of contemporary science:

> the historical distinctions between science and technology, pure and applied science, academic, industrial and military research, today hold weakly if at all. In the biotechnosciences, researchers move seamlessly between them all, as consultants, entrepreneurs, company directors and shareholders. Some could, with equal accuracy, be named capitalists as much as scientists.
>
> (Rose and Rose 2012: 11)

This implicates new sites and subjects into what has elsewhere been termed a new era or domain of capitalism: biocapital, or the maintenance and enhancement of 'life' as a political and economic project. This has also led to the inducement to scientists to approach their work as entrepreneurs and for human subjects participating in biomedical research to view their involvement as a form of 'work on the self' (Cooper and Waldby 2014).

In the 'bioeconomy', it is no longer sufficient for biological and medical researchers to pursue knowledge as an end in itself. Rather, post-war biology must direct its pursuits towards the application of biological knowledge for some human benefit – to pursue health, rather than truth (Rabinow 1996; Rabinow and Rose 2006). The alignment of public institutions like universities with capitalist enterprises through competitive mechanisms of governance, regimes of accountability, and other forms of 'audit culture' is both a defining feature of the knowledge economy and a characteristic of contemporary life sciences and medical research. These fields are seen as sites for investing venture capital, creating new forms of public–private partnership, and supporting spin-offs from publicly funded research into for-profit enterprises. An uneven landscape of scientific change, partial regulatory oversight, and the framing of what is within and outside of the bioethical frame call for close attention to the new subjects and spaces of the life sciences, medicine, and health.

Liberal Euro-American conceptions of the body as property enable the transformation of bodily materials into new kinds of objects. These objects are described as 'resources', 'scientific tools', 'technologies', or even new kinds of 'assets'. Technologies of preservation, transformation, mobility, and exchange are enabling new geographies of bodily materials to emerge, one of the defining characteristics of what commentators like anthropologist Kaushik Sunder Rajan (2006) call a new era of 'biocapital'. Work attuned to the subjective dimensions of participation in the bioeconomy, for example, demonstrates that the desire to see value in one's bodily contribution takes different forms very much dependent on an increasingly global social field, as bodies and body parts, medical practitioners, and patients seeking treatment move across borders and around the globe (Parry *et al.* 2015). Detailed ethnographic accounts attest to the existence of flourishing markets for certain tissues, including eggs (Ikemoto 2009; Nahman 2008; Widdows 2009) and kidneys (Cohen 2003). There is no consensus, for example, on whether payment for individuals who provide tissues for research should be regarded as an undue inducement, a form of compensation, a wage, or something else (Waldby *et al.* 2013). The exchange of bodily parts, the implicit or explicit markets for organs and tissues, and identification of biological materials formerly discarded as waste as new 'resources' for research or therapy thus map new geographies of economic unevenness.

There are also ways in which reflections on the political economy of the life sciences, 'biocapital', and the 'bioeconomy' pose new questions about sharing in common. This chapter also considers how the turn to 'commons' thinking and practice in the life sciences reveals what some commentators identify as the evolution of 'biocapital' towards new forms of sharing and the pooling of resources or 'assets' (Lezaun and Montgomery 2015; Rajan 2006). Indeed, the practices and discourses of 'sharing' are increasingly becoming part of the repertoire of partnerships between private and public entities. Javier Lezaun and Catherine M. Montgomery (2015) consider the emergence of product development partnerships in the pharmaceutical industry as an important site of transformation in the relationship between property and value towards what they name a 'pharmaceutical commons': whereas protective mechanisms like intellectual property regimes were once deployed primarily to defend exclusive use of knowledge related to product development, the pharmaceutical industry is increasingly engaged in processes of 'commoning' where intellectual property is used as an incentive to collaboration between researchers and product developers. Sharing knowledge, materials, and an ethos of partnering and 'open innovation' are now imperatives in the research and development arms of the pharmaceutical industry. These processes suggest that commoning is also emerging as a new way of governing the enclosures of knowledge at work in the bioeconomy.

Despite the rhetorical force of claims that the contemporary bioeconomy is characterised by 'bodies for sale' (Scheper-Hughes 2001), the creation of markets for bodies and body parts is only one part of the story of how biological materials become objects of circulation and exchange. This chapter concludes with a consideration of the extent to which biobanking initiatives are identified as sites for the creation of new bodily commons. It considers how biobanks are seen to threaten, devalue, or shrink an expansive notion of a 'bodily commons' or alternatively, constitute a commons in new ways. This discussion draws on empirical research carried out over several years on the creation of biobanks in the USA and UK, primarily those aimed at storing tissues for personal therapy but also as a 'public' resource for research use. The focus of this research is on biobanks that specifically collect tissues associated with pregnancy and where maternity and the processes of human growth and development *in utero* are key sites for the collection and generation of biological materials.

The following questions guide this chapter's consideration of a bodily commons: first, what are the contours of the political horizon of a bodily commons? How is this commons shrinking under the threat of privatisation, commodification, enclosure, and dispossession? Or is the political horizon of a bodily commons engendering new forms of sharing and other critical resources for constructing the subjects and spaces of the bioeconomy in new and more progressive ways? Although we may not know the eventual fate of this commons, what resources might we draw on to develop more critical approaches to it? This is increasingly the question asked as 'commons' thinking makes its way from the legacy of Elinor Ostrom to arrive at recent reflections on the management and theory of biobanks as commons (Huzair and Papioannou 2012). These reflections on the commons are informed by feminist and autonomist Marxist accounts of the commons 'as the terrain on which to promote, and struggle for, different processes of valorisation' (Allesandrini 2012: 5).

Genetic commons: the body encoded

One of the most frequent figures of debate over bodily commons is that of the body's genetic 'code'. Here, the commons is imagined as a kind of 'shared genetic heritage' that cannot be owned outright because it is believed to be the result of millennia of evolutionary forces. As may well be familiar, the efforts to identify all the genetic sequences of the human body capitalised on and further intensified investment in techniques to interpret genetic material for research and potential therapeutic use. This intensification of investment, hope, and the language of 'discovery' helped solidify bioprospecting in the public and scientific imagination as spin-off endeavours, such as National Geographic's Genographic Project, sought to map and track the genetic diversity of a shared human or species heritage, all while making presumptions about a human identity in common – invoked by the phrase 'We are all African' that obscures historical and social hierarchies and inequities (Nash 2015).

As Gísli Pálsson and Barbara Prainsack's discussion of 'genomic stuff' highlights, there are several dimensions to the notion of a genetic commons in today's *post*-genomic era. The rhetorical claims to a common human heritage are enshrined in the doctrine of the Universal Declaration on the Human Genome and Human Rights, insofar as 'the genome "underlies the fundamental unity of all members of the human family.... In a symbolic sense, it is the heritage of humanity"' (Pálsson and Prainsack 2011: 277, citing UNESCO 1997). Pálsson and Prainsack's exploration of the notion of a genetic commons consider the governance of genetic materials and the bioscientific technologies of genetic and genomic research as an important contribution to thinking the future of common property regimes. Traditional notions of common resources are inadequate for understanding the new socio-technical objects that they name 'genomic stuff'. They coin this neologism as a way to signal the overlapping constitution of material and immaterial elements in the making of the 'gene'. The gene is a complex hybrid of data, information, and material. No meaning is inherent in 'it'. As a research object, it invites renewed consideration of the presumed boundaries between material object and immaterial meaning. Its geography is also at issue: materials stored in a discrete location may be transformed into information on a database and exchanged globally. In addition to the call by Melinda Cooper and Catherine Waldby (2014) to recognise the contribution of experimental subjects and tissue donors as a form of 'clinical labour', Pálsson and Prainsack argue that the transformation of saliva, blood, or other bodily materials into information in transnational biobanks or databases makes the 'providers of genomic stuff ... "virtual migrants"' in which 'the "same" body, then, in a sense, performs labour at two or more sites simultaneously' (2011: 266, citing Aneesh 2006).

The labour of the gene or genetic body as a hybrid of information, data, and material for understanding disease processes and etiologies is further complicated by the suggestion that genes no longer carry the powers of causation once invested in them. Epigenetics instead suggests that environmental factors once

seen as external to genetic forms of inheritance are much more complexly implicated in the way environmental forces leave a 'trace' within the body, a pattern
of genetic expression that can be inherited and are thus suggestive of a much
closer relationship between genes and environment than previously understood.
Indeed, if one were to rework how the genetic commons differs from other kinds
of commons, or is remade by the very technologies that instantiate it, it would be
to see epigenetics as the principle that folds the outside of the gene within the
very architecture of genetic identity. The body, after epigenetics, is increasingly
conceived as a porous entity, ontologically made *with* the environment (Pálsson
and Prainsack 2011; see also Meloni and Testa 2014).

Alongside this *epi*genetic commons, microbiome projects signal another
remaking of the notion of a genetic commons: here the exercise of genome
mapping does not affirm the human body as self-identical but as radically other
to its 'self'. The body is host to teeming populations of microbial life, varied and
diverse, now mappable in new geographies – for example, in artist Rebecca D.
Harris's work 'Symbiosis', a two-dimensional outline of the pregnant human
body is embellished with thousands of French knots, each representing the distribution of microbes and shading from blue to pink to red and orange to make
visual the diverse ecologies that undulate across the exterior surface of the skin.
Harris also makes visible the interior surface of the uterus (see Colls and Fannin
2013) and leaves this space blank, the unknotted surface of the canvas around
the foetus representing the long-held assumption that the uterus is a 'sterile'
environment: Harris writes of this 'blank' interior space that the foetus 'quietly
awaits its journey into the world for its first seeding of microbes' (Harris 2015).
As poignant as this representation of the interior bodily space of the body is, it
now depicts an anachronistic understanding of uterine life. Microbiological and
immunological research suggests that even this interior bodily space hosts its
own microbiome (Aargaard *et al.* 2014; Prince *et al.* 2014). 'Big Science' projects like the NIH Human Microbiome Project supported by the US National
Institute of Health take this expanded notion of a bodily commons as a starting
point, where what is shared between human bodies is a genetic as well as microbial commons or number of 'communities'. The institutional and political
economic dimensions of such large-scale projects rely on the dynamics of collaboration and competition that shaped previous multi-sited, global projects to
map the human genome.

The language of the 'commons' proliferates in reflections on the problems of
property, labour, and value in the bioeconomy: the 'genetic commons' may
refer to the moral injunction to envision the gene pool of all life on the planet as
a 'global commons' (Treaty to Share Genetic Commons 2002), or to national
initiatives to collect resources from discrete laboratories and biobanks into build
a common platform for research (Huzair and Papaioannou 2012). Deepa S.
Reddy terms the biomedical commons: 'a space of common resources cautiously guarded against commercial encroachment: the legally regulated public
domain of access, sharing and innovation' (2013: 283). Indeed, invoking
the language of the commons may create possibilities for a new forms of

participation and benefit-sharing (Oldham 2009). It is on this last conception of the commons that the following discussion will focus by drawing out the significance of the spaces and times of life as central to new enclosures and openings in a 'bodily commons'.

Expropriation and the laboring body

Michael Hardt cites the domains of 'patents, copyrights, indigenous knowledges, genetic codes, the information in the germplasm as seeds' (2010: 349) as the terrain over which new problems of the commons are being posed and debated. In these arenas of debate, scarcity, Hardt argues, is no longer the underlying or dominant economic logic. The sharing of genetic knowledge between researchers, for example, actually increases, rather than decreases, the value of these entities as immaterial goods. Knowledge is an immaterial aspect of the genetic or information commons that Hardt suggests is immanent to capital, where the logic of property that underpins forms of private or state ownership cannot keep up with the 'pressure to ... become common' (Hardt 2010: 349). Hardt places efforts to patent the information gleaned from genetic sequencing technologies as part of longer and more extensive forms of expropriation and 'primitive accumulation' processes by which capital attempts to turn the common into property. Simultaneous to this expropriation, however, a commons is constantly produced and projected out of the radically open and contingent potential of all (social) production. The forms of immaterial production that Hardt identifies as the dominant sites of capital accumulation are the production of information, affects, as well as the biopolitical production of humans as living, sensing, and feeling social beings. One of the defining tasks at hand, then, in an investigation of the commons is to consider the subjectifying dimensions of what Hardt terms 'class composition – asking, in other words, how people produce, what they produce, and under what conditions, both in and outside the workplace, both in and outside relations of wage labour' (2010: 355).

Hardt presumes a kind of ontological primacy of the 'commons' as such; the ongoing *production of the common* as a process is precisely what capital seeks to appropriate, contain, and harness for its own ends. In affirming this immanent production of the common as a continual process inherent to capitalist dynamics of enclosure, he imagines a future horizon for communism, albeit not one in which private forms of property are abolished only to be replaced by public, shared, or 'universal' relations of property. Rather, 'communism properly conceived instead is the abolition of not only of private property but of property as such' (Hardt 2010: 352).

In contrast to Hardt's epochal notion of the expansion of 'immaterial' labour as what distinguishes contemporary capitalism from its previous forms, feminist scholars of the political economies of the biosciences signal the ongoing transformation of the material relations of production and social reproduction (Cooper and Waldby 2014; Dickenson 2007; Kent 2008; Waldby and Cooper 2008, 2010). Highlighting the domains in which the fleshy body becomes subject to regimes of intellectual property, this work tracks how the body is fragmented, disaggregated, separated, and parsed out into moveable and potentially exchangeable parts from which it is possible to discern their transformation into commodities, and the creation of new forms of research and market value. Parts of the body previously devalued as waste become newly valued as research resources, cell lines patented and further commodified, reproduced on an industrial scale and sold on a market (Landecker 2000; Waldby and Mitchell 2006). The body is presumed to be *res nullius*, a space that belongs to no one. To stake a claim of property in one's own body is to objectify what makes up one's personhood, one's relationship as a living subject to the body one is. Thus the moral injunction that the body first and foremost belongs to no one has indeed made possible the claims to property for those whose intellectual labour transforms the material of the body into knowledge, innovation, tool, and technology.

Melinda Cooper and Catherine Waldby seek to shift the terms of feminist debate over property in the body by moving from a liberal political fiction of the autonomous subject, in which life is conceived as an intrinsic property of a living body and the ideal relationship of subject-body is imagined as one of ownership, to a consideration of the temporal unfolding of development, growth, and change and the 'enclosure' of this process in the bioeconomy. Cooper and Waldby describe the attempt to capture, enclose, and control this temporal unfolding as a form of 'clinical labor ... the process of *material abstraction* by which the abstract, temporal imperatives of the accumulation are put to work at the level of the body' (Cooper and Waldby 2014: 12, emphasis in original). Clinical labour is *not*, in their theorisation, a way to describe the process by which capitalism encloses a kind of essential vitality or 'intrinsic generativity of living biology' but rather a way to 'render the biological newly pliable to the exigencies of abstract, exchangeable, or statistical time' (Cooper and Waldby 2014: 12). What is captured is not a notion of the material as *essence*, but the materiality of becoming, the durational dimension of time and the transformation of heterogenous bodily difference into measurable units. This compels the question of what an affirmative bodily commons would require: what is done with the time of the living body? What is shared with this time? What space given to growth, development, and the unfolding of life processes? To what extent is biological being – or the bodily commons of a new kind of species-being – rendered in new ways compatible with or resistant to the abstractions and temporalisations of capitalist political economy? Experiments to collectivise or make more communal the benefits of scientific research are under way, yet there remains considerable work to be done to trace the emerging practices, subjects, and spaces of recruitment related to the outsourcing of clinical trials, of nascent forms of women's activism as clinical labourers (DasGupta and Dasgupta 2014), and of evolving debates over the sharing of benefits from the growth of biobanks (Hayden 2007; Parry and Gere 2006). These initiatives and experiments signal an alternative horizon for a 'bodily commons'.

Corporealising a commons? Biobanking as site of enclosures and openings

In the post-genomic era, the collection of bodily materials from individuals and populations, and the effort to link these collections with a range of medical, behavioural, and genealogical data, have increased. Efforts to account for the intensification of the collection of biological materials are made difficult by the diffuse and dispersed nature of collections: local, regional, and national biobanks may be involved in more or less coordinated initiatives to standardise material and data collection and storage. The presumptions underlying nation-state centred gift regimes involving organs and blood for transplantation and transfusion remain salient but are supplemented by utilitarian arguments to increase donations of blood and tissues for transplantation (Clarke *et al.* 2003; Sothern and Dickinson 2011). This utilitarianism is reflected in contemporary biobanking initiatives in which bodily materials are framed as deserving of 'efficient' or 'wise' use, and which involve identifying novel uses for waste materials such as umbilical cord blood (Brown and Kraft 2006; Brown *et al.* 2011; Fannin 2011), aborted foetal material (Kent 2008; Pfeffer 2009), menstrual blood (Fannin 2013), placenta (Fannin and Kent 2015), and other tissues.

Herbert Gottweis and Alan Petersen frame the growth of biobanking as a strategic concern in the biosciences, with social, ethical, economic, and political implications for the communities of scientists, curators, and participants involved in establishing a biobank (Gottweis and Petersen 2008). They emphasise that much of the public and policy discourse concerning biobanks has tended to focus on the notion of 'risk' for the individual participant, insofar as the key bioethical issues relate to notions of informed consent, ensuring the dignity of participants, confidentiality and other concerns that reflect efforts to protect the privacy and rights of self-determination of donors. Debates over the governance of biobanks are thus concerned, at the level of policy, regulation, and bioethical deliberation 'with the goal to integrate biobanks into the existing fabric of regulation, medicine, law and society' (Gottweis and Petersen 2008: 7). Much of the discussion of the ethics and regulation of biobanks is oriented towards a consideration of the differential rights of donors/patients, and the intellectual property rights of researchers, rather than the political economies of science (Gottweis and Petersen 2008; Høyer 2002, 2008). The ethical dilemmas posed by biobanks tend to be framed as how best to protect the rights-bearing individual, rather than to open up the question of the creation of a bodily commons to greater interrogation.

Yet biobanks are also new forms of governing life. The sharing of data, resources, open source, and open access are all part of an emerging configuration of biobanks as bodily commons and of commoning as one of the techniques of new forms of governing. Early reflections on the governance of biobanking in the field of bioethics were focused on the protection of the rights of the participants (rights to privacy, informed consent, and so on) as well as the protection of the rights of researchers to gain property rights over their 'inventions' and innovations as they transformed collected materials into cell lines, products, and technical resources. Today, these considerations no longer dominate debate over the ethics of biobanking; rather, the emphasis is on economic logics of scarcity and the maximisation of value from collected samples in light of the experimental and professional challenges related to the maintenance and use of banked materials (Tupasela 2011). Technical considerations over the exchangeability of materials in biobanks thus tend to dominate, and challenges to the presumed social value of biobanking for health care and research seem to find less secure footing.

Indeed, ambiguities over how to attribute ownership to a tissue are multiplied when the tissue in question is produced 'in common'. Placenta, umbilical cord blood, and other tissues that develop during pregnancy come into being through shared biological processes of growth and development. The 'identity' of these biological materials may be accounted for and named in different ways. The genetic composition of placental material makes it 'foetal', but the identity of the person making a legal claim to its use and disposition after birth is not the child but may be one of many others. Indeed, when asked about the identity of the placenta in a recent study with mothers who contributed to a placenta biobank, ambiguities abound. Rather than seeking to establish property ownership or identity as a contest between competing interests or autonomous genetic or biological subjects, the placenta biobank sought to enhance the value of the collected placentas by retaining them as 'connective tissues' that gained scientific value through the 'fixing' of maternal/foetal relations in time and space (Fannin and Kent 2015).

The enclosure of the bodily commons of development and growth in a biobank seeks to arrest time, freezing or fixing the temporal degradation of a living tissue in order to make it available for potential future use. Viewed in this way, the future horizon of a biobanking bodily commons materialises the profound desire to 'hoard' the generative capacities of living beings in order to construct a future biological and species-being (Fannin 2013). Here, the notion of a bodily commons draws attention to the specificities of a collection and to the way a commons also constitutes a community. The etymological connection between 'commons' and 'community' signals how the bringing together of material things also implicates the creation of new relationships and new subjects. The materials collected in biobanks thus represent not just one commons but many, a commons of technologies, practices, discourses, and relations around 'the future human'.

Futures for a bodily commons

How might a bodily commons come into being through the creation of new scientific objects and research tools *and* in relation to other notions of a 'commons of the body' and other aspects of the 'human sensorium': 'seeing, hearing, smelling, tasting, feeling thinking, contemplating, sensing, wanting, acting, loving' (Hardt 2010: 353, citing Marx 1975: 351). How might work on the commons of the body be a space for the cultivation of desire or for the flourishing of a new sensorium in relation to living being? This requires care and vigilance to guard against the creation of a 'commons of the body' that configures some bodies as more available and open for exploitation while also attending to the bodily differences that subtend this configuration.

The identification of ever new ways in which bodily materials can be transformed into scientific objects recalls Marx's invocation of the ceaseless overturning of old relations of production and 'old wants' into ever more restless, cosmopolitan desires for the new. Viewed in this way, the production of a scientific object is also always the production of new relationships and new subjects (Anderson 2013; see also Strathern 1992). The subjective and processual dimensions of the making of new objects – and new subjects – of the bioeconomy are intimately linked to the new kinds of commons generated by the collection and exchange of bodily materials. Reflecting on the power and limits of the metaphor of enclosure, Donna Haraway asserts:

> The figure we often used to tell the story of commodification is an enclosure of a commons, but it isn't enough. For example, genomes are not being enclosed (or not only being enclosed); they are coming to be out of the action of many players, human and not. Genomes are generating new kinds of wealth and, as Sarah Franklin and Margaret Lock (2003) put it, new ways of living and dying. Enclosure is too narrow a metaphor. You can't understand technobiocapital through 18th century agricultural commodification. There is a whole lot going on besides enclosure.
>
> (Donna Haraway, in Gane 2006: 149)

Besides enclosure, scientific objects may be increasingly valued under new logics of scarcity, wise use, or the ethos of a sharing economy. The bodily commons of biobanks also generate biopolitical community: 'affective networks' of researchers, participants, institutions, media, and banked samples, a commons and community whose futures may be unknown (see Fannin and Kent 2015; Kowal 2013).

The spaces of the body are not often imagined, in Euro-American liberal legal or political culture, as 'common' spaces but indeed this is precisely what led Richard Titmuss to develop his concept of the social 'right to share' from the blood donation system. Although the notion of a commons is perhaps much more frequently debated and understood as a dimension of how land, water, air, or other material entities can be managed, Titmuss' consideration of the blood donation system drew on the notion that certain entities were inappropriable because they could not or should not belong to any one person or people, or even species. Titmuss' landmark text (1971) on the social policy implications of blood donation systems in the UK and USA was written in the wake of 'common property' resource debates in the 1950s over how to grapple with the elemental fluidity of water and air, capable of traversing the bounded territories constituted by legal regimes and modern forms of political sovereignty. This elemental fluidity of air and water, in contrast to conceptions of the earth as a bounded and bordered surface, continues to serve as a critical resource for thinking about space and territory anew.

This chapter has examined the fluid space 'closest in' – the commons of bodily becoming, the processes of growth and development proper to living bodies, and the material and relational configurations generated from these bodies. Broad reflections on the enclosures of neoliberal capitalism do not always fully incorporate the significant transformations of labour, property, and value occurring in the domain of biocapital and the bioeconomy. The life sciences, medicine, and health are characterised by shifting boundaries between what is understood to be a commodity or commodifiable and what is held in common or as a commons. The social dynamics of health, kinship, and community, and the distribution of life risks between populations shape and are shaped by new enclosures and openings in a bodily commons.

Invoking this bodily commons speaks to an alternative set of orientations to desire, to notions of sharing and communality that liberate rather than constrain. The appeal to the 'common' is an appeal to a different orientation to the world, to living in the world, and to a new horizon for politics (Hardt 2010). Thinking through what the commons of the body might be and what future horizon a bodily commons opens up is therefore a necessary and urgent task.

References

Aargaard, Kjersti, Jun Ma, Kathleen M. Antony, Radhika Ganu, Joseph Petrosino, and James Versalovic. 'The Placenta Harbors a Unique Microbiome'. *Science Translational Medicine* 6, no. 237ra65 (2014), at: http://dx.doi.org/10.1126/scitranslmed.3008599 (accessed 18 January 2016).

Alessandrini, Donatella. 'Immaterial Labour and Alternative Valorisation Processes in Italian Feminist Debates: (Re)exploring the "Commons" of Re-production'. *feminists@ law* 1, no. 2 (2012), at: http://journals.kent.ac.uk/index.php/feministsatlaw/article/view/32 (accessed 18 January 2016).

Anderson, Warwick. 'Objectivity and Its Discontents'. *Social Studies of Science* 43, no. 4 (2013): 557–76.

Aneesh, A. *Virtual Migration: The Programming of Globalization.* Durham, NC, Duke University Press, 2006.

Boyle, James. 'The Second Enclosure Movement and the Construction of the Public Domain'. *Law and Contemporary Problems* 66 (2003): 33–74.

Brown, Nik and Alison Kraft. 'Blood Ties: Banking the Stem Cell Promise'. *Technology Analysis and Strategic Management* 18, no. 3/4 (2006): 313–27.

Brown, Nik, Laura L. Machin, and Dana McLeod. 'Immunitary Bioeconomy: The Economisation of Life in the International Cord Blood Market'. *Social Science & Medicine* 72, no. 7 (2011): 1115–22.

Chakrabarty, Dipesh. 'The Climate of History: Four Theses'. *Critical Inquiry* 35, no. 2 (2009): 197–222.

Clarke, Adele E., Janet K. Shim, Laura Mamo, Jennifer Ruth Fosket, and Jennifer R. Fishman. 'Biomedicalization: Technoscientific Transformations of Health, Illness, and U.S. Biomedicine'. *American Sociological Review* 68, no. 2 (2003): 161–94.

Cohen, Lawrence. 'Where It Hurts: Indian Material for an Ethics of Organ Transplantation'. *Zygon* 38, no. 3 (2003): 663–88.

Colls, Rachel and Maria Fannin. 'Placental Surfaces and the Geographies of Bodily Interiors.' *Environment and Planning A* 45.5 (2013): 1087–104.

Cooper, Melinda and Catherine Waldby. *Clinical Labour: Tissue Donors and Research Subjects in the Global Bioeconomy*. Durham, NC: Duke University Press, 2014.

DasGupta, Sayantani and Dasgupta, Shamita Das (eds). *Globalization and Transnational Surrogacy in India: Outsourcing Life*. Lanham, MD: Lexington Books, 2014.

Dickenson, Donna. *Property in the Body: Feminist Perspectives*. Cambridge and New York: Cambridge University Press, 2007.

Fannin, Maria. 'Personal Stem Cell Banking and the Problem with Property'. *Social & Cultural Geography* 12, no. 4 (2011): 339–56.

Fannin, Maria. 'The Hoarding Economy of Endometrial Stem Cell Storage'. *Body & Society* 19, no. 4 (2013): 32–60.

Fannin, Maria and Julie Kent. 'Origin Stories from a Regional Placenta Tissue Collection'. *New Genetics and Society* 34, no. 1 (2015): 25–51.

Franklin, Sarah and Margaret Lock (eds). *Remaking Life and Death*. Santa Fe, NM: School of American Research, 2003.

Gane, Nicholas. 'When We Have Never Been Human, What Is to Be Done? Interview with Donna Haraway'. *Theory, Culture & Society* 23, no. 7–8 (2006): 135–58.

Gottweis, Herbert and Alan Petersen (eds). *Biobanks: Governance in Comparative Perspective*. Abingdon: Routledge, 2008.

Haraway, Donna. 'Deanimations: Maps and Portraits of Life Itself', in Avtar Brah and Annie Coombes (eds), *Hybridity and Its Discontents: Politics, Science, Culture*. New York, Routledge, 2005, 111–36.

Hardt, Michael. 'The Common in Communism'. *Rethinking Marxism* 22, no. 3 (2010): 346–56.

Harris, Rebecca D. 'Portfolio'. 2015, at: www.rebecca-harris.com/#!symbiosis/c1ips (accessed 27 August 2015).

Hayden, Cori. 'Taking as Giving Bioscience, Exchange, and the Politics of Benefit-sharing'. *Social Studies of Science* 37, no. 5 (2007): 729–58.

Høyer, Klaus. 'Conflicting Notions of Personhood in Genetic Research'. *Anthropology Today* 18, no. 5 (2002): 9–13.

Høyer, Klaus. 'The Ethics of Research Biobanking: A Critical Review of the Literature'. *Biotechnology and Genetic Engineering Reviews* 25 (2008): 429–52.

Huzair, Farah and Theo Papaioannou. 'UK Biobank: Consequences for Commons and Innovation'. *Science and Public Policy* 39, no. 4 (2012): 500–12.

Ikemoto, Lisa C. 'Eggs as Capital: Human Egg Procurement in the Fertility Industry and the Stem Cell Research Enterprise'. *Signs: Journal of Women in Culture and Society* 34, no. 4 (2009): 763–81.

Kent, Julie. 'The Fetal Tissue Economy: From the Abortion Clinic to the Stem Cell Laboratory'. *Social Science & Medicine* 67, no. 11 (2008): 1747–56.

Kowal, Emma. 'Orphan DNA: Indigenous Samples, Ethical Biovalue and Postcolonial Science'. *Social Studies of Science* 43, no. 4 (2013): 577–97.

Landecker, Hannah. 'Immortality, In Vitro: History of the HeLa Cell Line', in Paul E. Brodwin (ed.), *Biotechnology and Culture: Bodies, Anxieties, Ethics*. Indianapolis, IN: Indiana University Press, 2000, 53–72.

Lezaun, Javier and Catherine M. Montgomery. 'The Pharmaceutical Commons: Sharing and Exclusion in Global Health Drug Development'. *Science, Technology & Human Values* 40, no. 1 (2015): 3–29.

Martin, Paul, Nik Brown, and Andrew Turner. 'Capitalizing Hope: The Commercial Development of Umbilical Cord Blood Banking'. *New Genetics and* Society 27, no. 2 (2008): 127–43.

Marx, Karl. *Economic and Philosophical Manuscripts* in *Early Writings*, trans. Rodney Livingstone and Gregor Benton. London: Penguin, 1975.

Meloni, Maurizio and Testa, Giuseppe. 'Scrutinizing the Epigenetics Revolution'. *BioSocieties* 9, no. 4 (2014): 431–56.

Nahman, Michal. 'Nodes of Desire: Romanian Egg Sellers, "Dignity" and Feminist Alliances in Transnational Ova Exchanges'. *European Journal of Women's Studies* 15, no. 2 (2008): 65–82.

Nash, Catherine. *Genetic Geographies*. Minneapolis, MN: University of Minnesota Press, 2015.

Oldham, Paul D. 'An Access and Benefit-Sharing Commons? The Role of Commons/ Open Source Licences in the International Regime on Access to Genetic Resources and Benefit-Sharing (July 23, 2009)'. *Initiative for the Prevention of Biopiracy, Research Documents*, Year IV, No. 11. http://dx.doi.org/10.2139/ssrn.1438027 (accessed 18 January 2016).

Pálsson, Gísli and Barbara Prainsack. 'Genomic Stuff: Governing the (Im)matter of Life'. *International Journal of the Commons* 5, no. 2 (2011): 259–83.

Parry, Bronwyn and Cathy Gere. 'Contested Bodies: Property Models and the Commodification of Human Biological Artefacts'. *Science as Culture* 15, no. 2 (2006): 139–58.

Parry, Bronwyn, Beth Greenhough, Tim Brown, and Isabel Dyck (eds). *Bodies Across Borders: The Global Circulation of Body Parts, Medical Tourists and Professionals*. Farnham: Ashgate Publishing, 2015.

Pfeffer, Naomi. 'How Work Reconfigures an "Unwanted" Pregnancy into "The Right Tool for the Job" in Stem Cell Research'. *Sociology of Health & Illness* 31 (2009): 98–111.

Prince, Amanda L., Kathleen M. Antony, Derrick M. Chu, and Kjersti M. Aagaard. 'The Microbiome, Parturition, and Timing of Birth: More Questions than Answers'. *Journal of Reproductive Immunology* 104 (2014): 12–19.

Rabinow, Paul. *Essays on the Anthropology of Reason*. Princeton, NJ: Princeton University Press, 1996.

Rabinow, Paul and Nikolas Rose. 'Biopower Today'. *BioSocieties* 1 (2006): 195–217.

Rajan, Kaushik S. *Biocapital: The Constitution of Postgenomic Life*. Durham NC: Duke University Press, 2006.

Reddy, Deepa S. 'Citizens in the Commons: Blood and Genetics in the Making of the Civic'. *Contemporary South Asia* 21, no. 3 (2013): 275–90.

Rose, Hilary and Steven Rose. *Genes, Cells and Brains: The Promethean Promises of the New Biology*. London: Verso, 2013.

Scheper-Hughes, Nancy. 'Bodies for Sale – Whole or in Parts'. *Body & Society* 7, no. 2 (2001): 1–8.

Sevilla-Buitrago, Alvaro. 'Capitalist Formations of Enclosure: Space and the Extinction of the Commons'. *Antipode* (2015) Ahead-of-print publication 30 July 2015, doi:10.1111/ anti.12143.

Sothern, Matthew and Jen Dickinson. 'Repaying the Gift of Life: Self-help, Organ Transfer and the Debt of Care'. *Social & Cultural Geography* 12, no. 8 (2011): 889–903.

Strathern, Marilyn. *Reproducing the Future: Anthropology, Kinship and the New Reproductive Technologies*. Manchester: Manchester University Press, 1992.

Titmuss, Richard. *The Gift Relationship: From Human Blood to Social Policy*. New York: Pantheon Books, 1971.

Treaty to Share Genetic Commons, 2002, at: www.foet.org/past/Treaty%20Document%20 English.html (accessed 20 August 2015).

Tupasela, Aaro. 'From Gift to Waste: Changing Policies in Biobanking Practices'. *Science and Public Policy* 38, no. 7 (2011): 510–20.

UNESCO *Universal Declaration on the Human Genome and Human Rights*. Paris: UNESCO, 1997.

Waldby, Catherine and Melinda Cooper. 'The Biopolitics of Reproduction: Post-Fordist Biotechnology and Women's Clinical Labour'. *Australian Feminist Studies* 23, no. 55 (2008): 57–73.

Waldby, Catherine and Melinda Cooper. 'From Reproductive Work to Regenerative Labour'. *Feminist Theory* 11, no. 1 (2010): 3–22.

Waldby, Catherine, Ian Kerridge, Margaret Boulos, and Katherine Carroll. 'From Altruism to Monetisation: Australian Women's Ideas about Money, Ethics and Research Eggs'. *Social Science & Medicine* 94 (2013): 34–42.

Waldby, Catherine and Robert Mitchell. *Tissue Economies: Blood, Organs, and Cell Lines in Late Capitalism*. Durham, NC: Duke University Press, 2006.

Widdows, Heather. 'Border Disputes Across Bodies: Exploitation in Trafficking for Prostitution and Egg Sale for Stem Cell Research'. *International Journal of Feminist Approaches to Bioethics* 2, no. 1 (2009): 5–24.

12 Commoning as a postcapitalist politics[1]

J.K. Gibson-Graham, Jenny Cameron, and Stephen Healy

Introduction

'The tragedy of the commons' is a well-known phrase that has captured people's imaginations for generations. Unfortunately these few but powerful words have been used to justify the enclosure and erasure of many well-functioning commons that benefit both people and the environment. Less well-known is Garrett Hardin's qualification in 1998 – some thirty years after he coined the original phrase – that:

> To judge from the critical literature, the weightiest mistake in my synthesizing paper was the omission of the modifying adjective 'unmanaged'.
>
> (Hardin 1998: 682)

What Hardin had presented in his original work was an open access and unmanaged pasture where there was no community that cared for the fields, took responsibility for them, organised herder access, negotiated grazing use and oversaw the distribution of benefit to community members (1968). It would be fair to say that Hardin's pasture bears little resemblance to the commons that researchers such as Elinor Ostrom (e.g. 1990) have meticulously documented, commons that have rules or protocols for access and use, and are cared for by a community which takes responsibility for the commons and distributes the benefits.

Today the planet faces a genuine tragedy of the unmanaged 'commons'. For decades an open access and unmanaged resource has been treated with the same sort of disregard as Hardin's pasture was treated. The planet's life-supporting atmosphere has been spoiled by ' "help yourself" or "feel free" attitudes' (Hardin 1998: 683). We are now faced with the seemingly impossible task of transforming an open access and unmanaged planetary resource into a commons which is managed and cared for. With the cause and impacts of global warming now beyond debate, we are being pressed to take responsibility and to act in new ways. But how are we to do this? What type of politics is called for?

In this chapter we explore how the process of commoning offers a politics for the Anthropocene. To reveal the political potential of commoning, however, we need to step outside of the ways that the commons have generally been understood. One predominant framing positions the commons in relation to capitalism, as Kevin St Martin writes:

> It would seem that all of our stories of the commons revolve around a capitalist imaginary: capitalism's origin in the enclosure of the commons, capitalism's commodification of natural resources, capitalism's expansion and its penetration of common property regimes globally, and capitalism's most recent push to privatize remaining common property resources via neoliberal policies at a variety of scales.
>
> (St Martin 2005: 63)

We discuss this capitalocentric framing of the commons in the first section and raise concerns about how this framing limits the potential of commoning as a politics for the Anthropocene. In the second section, we discuss a second predominant framing of commons as a 'thing' that is associated with publically owned or open access property. Instead we argue that commons can be conceived of as a process – commoning – that is applicable to any form of property, whether private, or state-owned, or open access. We then turn to three examples from the past and the present that provide insights into ways of commoning the atmosphere. We reveal how a politics of commoning has been enacted through assemblages comprised of social movements, technological advances, institutional arrangements, and non-human 'others'. In the final section, we discuss the implications of this understanding of politics and particularly what it means for understanding how transformation occurs.

Commons and capital

The term capitalocentrism was coined by Gibson-Graham (1996) to extend the feminist theorisation of phallogocentrism to the field of economy. Capitalocentrism names the way that a diversity of economic relations are positioned as either the same as, a complement to, the opposite of, subordinate to, or contained within 'capitalism'. The commons have been drawn into a capitalocentric discourse in discussions of what has been called 'the new enclosures' (De Angelis 2010: 03/17). The new enclosures builds on Marx's influential historical account in *Capital* Volume 1 of the land enclosures that accompanied the rise of capitalist agriculture and industry in Britain. In Marx's writing the historical events of clearance, enclosure, and legal capture of common property are distilled and abstracted into the theory of primitive accumulation. Subsequent Marxists have noted the 'need' of the capitalist system to sustain its extensive growth by continually finding and enclosing new commons, thus advancing an added dimension to the theory of capital accumulation by showing how so-called primitive accumulation is ongoing (De Angelis 2010; Harvey 2003). Rather than a linear movement from primitive to capitalist accumulation, researchers now talk of 'a constant back-and-forth movement in which primitive accumulation continually reappears and coexists with capitalist production' (Hardt and Negri 2009: 138).

One strand of work has focused on the contemporary enclosure (via privatisation) of the material and biophysical commons (e.g. land, forests, waters, and fisheries); a second has focused on the enclosure of immaterial commons (e.g. knowledges, languages, images, and codes). This second strand, which we discuss here in order to demonstrate the capitalocentrism of commons research, is associated with Michael Hardt and Antonio Negri's influential book *Commonwealth* (2009). With the rise of the networked information economy, Hardt and Negri are keen to highlight the central role that this immaterial commons, or what they call 'the common', now plays in capitalist accumulation. Think, for example, of advances in scientific knowledge such as genetic coding or biopiracy (the patenting of traditional ecological knowledge). Unlike the commons, 'the common does not lend itself to a logic of scarcity' (Hardt and Negri 2009: 146). Productivity increases not when the common is controlled and privatised but when it is shared and added to. Thus there is a cycle 'from the existing common to a new common, which in turn serves in the next moment of expanding production' (Hardt and Negri 2009: 145). For Hardt and Negri the seeds of liberation are to be found in this necessity for capital to share the common in order to increase productivity, while simultaneously also needing to enclose the common. In a later piece, Hardt describes this tension as providing 'the conditions and weapons for a communist project', one in which capital 'is creating its own gravediggers' (2010: 355). Massimo De Angelis and David Harvie similarly draw attention to this tension which they describe as: '[t]he "ambiguity" between commons-within-and-for-capital and commoning-beyond-capital' (2013: 291). For them, this is politically 'a razor edge that both capital and social movements must attempt to negotiate' (De Angelis and Harvie 2013: 291).

In these discussions of 'the new enclosures', we see practices of commoning drawn into a discourse that places capital at the gravitational centre of meaning making.[2] On the one hand there is much to commend in these strong formulations – not least being the sense of outrage that the enclosure analysis elicits and the elusive hopes of an emergent communism that the gravedigger account kindles. While the political sentiments of present outrage and hopes for future emancipation clearly retain mobilising force, we question the relevance of this mode of politics for the Anthropocene. Critical thinkers are suggesting that the scale of the climate crisis demands a different way of thinking about humans and human activity. The historian Dipesh Chakrabarty, for example, proposes that rather than thinking in terms of the history of capitalism and modernisation we need to take a deep history approach and consider ourselves as not just a species within the multi-species community of life on this planet but a species whose existence depends on other species (2009). Similarly, the philosopher Val Plumwood counselled that if our species is to survive the Anthropocene we need 'to imagine and work out new ways to live with the earth, to rework ourselves'; in other words we need to 'go onwards in a different mode of humanity, or not at all' (Plumwood 2007: 1). A politics grounded in capitalocentrism seems to offer little in the way of helping us to reposition ourselves for living on a climate changing planet. Might thinking about the commons and a politics of the commons outside the confines and strictures of capitalocentrism help us reimagine our ways of living on this planetary home?

Commons, commoning, and communities

Another way of thinking about commons has been to focus on it as a verb, as commoning. This formulation was explicitly introduced by Peter Linebaugh in his book, *The Magna Carta Manifesto: Liberties and Commons for All* (2008). He writes,

> To speak of the commons as if it were a natural resource is misleading at best and dangerous at worst – the commons is an activity and, if anything, it expresses relationships in society that are inseparable from relations to nature. It might be better to keep the word as a verb, an activity, rather than as a noun, a substantive.
>
> (Linebaugh 2008: 279)

We are drawn to Linebaugh's formulation of commons and commoning as expressing relationships between humans, and between humans and the world around.[3] In our recent book *Take Back the Economy: An Ethical Guide to Transforming our Communities* (2013), we characterise commoning as a relational process – or more often a struggle – of negotiating access, use, benefit, care, and responsibility (see Table 12.1). Commoning thus involves establishing rules or protocols for access and use, taking caring of and accepting responsibility for a resource, and distributing the benefits in ways that take into account the well-being of others. When these relationships are in place, what results are any number of commons including biophysical commons (e.g. soil, water, air, plant, and animal ecologies), cultural commons (e.g. language, musical heritage, sacred symbols, and artworks), social commons (e.g. educational, health, and political systems), and knowledge commons (e.g. Indigenous ecological knowledge, scientific, and technical knowledge). The resulting commons may also be of varying and overlapping scales from the household and family to the national and international; from the micro (such as a microclimate) to the macro (such as the planet's atmosphere).

Table 12.1 Commons negotiations

Access	Use	Benefit	Care	Responsibility	Property
Shared and wide	Managed by a community	Widely distributed to a community and beyond	Performed by community members	Assumed by a community	Any form of ownership (private, state, open access)

Adapted from Gibson-Graham *et al.* (2013).

One implication of this focus on commoning as a relational process is that it emphasises the role of communities in commoning. Listen, for example, to the anthropologist Stephen Gudeman who observes that '[w]ithout a commons, there is no community; without a community, there is no commons'; and that '[t]aking away the commons destroys community, and destroying a complex of relationships demolishes a commons' (Gudeman 2001: 27). For some, this idea of community is so tainted with nostalgia and romanticism that it is if not a dangerous concept then at least a naive one. But the community that commons is not pre-given; rather, communities are constituted through the process of commoning. As the geographer Amanda Huron puts it, '[t]here appears to be a dialectical relationship between commons formation and community formations: one does not necessarily precede the other' (Huron 2015: 370). This is not a straightforward or easy process. In line with the complexity of the task of commoning in the twenty-first century, the community that is assembled does not share an essence and may indeed comprise those who in other situations are locked in antagonistic relationships. For example, in *Take Back the Economy* we discuss a community that has gathered to create a commons to protect the endangered bridled nailtail wallaby (Gibson-Graham *et al.* 2013: 139–42). This community includes the unlikely mix of family farmers, concerned conservationists, sporting shooters, academic researchers, government rangers, beef cattle, a particular species of wallaby, and brigalow scrub (a form of remnant wooded grassland in western Queensland). All that this multi-species community shares is what Jean-Luc Nancy calls 'being-in-common, or being-with' (Nancy 1991: 2).

A second implication of this formulation vis-à-vis other formulations of the commons is that when we frame commons as an activity, or process, or practice rather than as a category we find that commoning can take place with any form of property, from privately owned property to open access property. As we summarise in Table 12.2, enclosed and unmanaged resources can be commoned not by changing ownership but by changing how access, use, benefit, care, and responsibility occur. In other words, ownership of property is largely a legal matter and need not deter resources from being commoned; or as Linebaugh puts it: 'Commoners think first not of title deeds, but of human deeds' (Linebaugh 2008: 45). As an example, the resources that are commoned to create a protected area for the bridled nailtail wallaby are based around the privately owned land of a farming family who have entered into a voluntary conservation agreement with the state government to protect one-fifth of their property from development, in perpetuity. But it also includes, for instance, the privately owned guns and bullets of the sporting shooters who kill the feral predators that threaten the wallabies, the state-owned cars and clipboards used by the government rangers to monitor the health of the wallabies; the collectively owned computers and open access internet used by the conservation groups who broadcast online to help protect the wallabies. What was an enclosed privately owned property (a portion of a family-owned farm) has been combined with a host of other resources of varying ownership to become a commons. Simultaneously, what we call a 'commoning-community' has been constituted and this community negotiates

Table 12.2 Ways of commoning

	Access	Use	Benefit	Care	Responsibility	Ownership
Commoning enclosed resources	Narrow	Restricted by owner	Private	Performed by owner or employee	Assumed by owner	Private individual Private collective State
Maintaining commons or creating new commons	Shared and wide	Managed by a community	Widely distributed to a community and beyond	Performed by community members	Assumed by a community	Private individual Private collective State Open access
Commoning unmanaged resources	Unrestricted	Open and unregulated	Finders keepers	None	None	Open access State

Adapted from Gibson-Graham *et al.* (2013).

access to the commons, determines how the commons are used and who bene-fits, and accepts care and responsibility (Table 12.2: row 1). Other instances of commoning enclosed property include the urban commons that are being created in central Dublin by communities who are using strategies such as fundraisers to pay rent for privately owned property so they can create what Bresnihan and Byrne describe as 'independent spaces' (2015: 36). Against the seeming tide of enclosures there is work to be done to understand more about how enclosed resources are being commoned. There is also work to be done to understand how unmanaged resources are being commoned (Table 12.2: row 3). This is where we turn our attention in the next section as we explore the tentative ways in which communities have engaged in commoning an unmanaged resource, the atmosphere.

But before we do this, it is worth returning for a moment to capitalocentric formulations of the commons, which have a tendency to characterise commons as principally a form of property. One expression of this is the homology in which private property is to capitalism, as state-owned property is to socialism, as common property is to primitive and future forms of communism (e.g. Hardt 2010: 346, 355). This thinking strategy privileges formal and abstract legalities at the expense of actual practices of maintaining or creating commons, or com-moning enclosed or unmanaged resources. In contrast, an anti-capitalocentric approach attends to the diversity of practices for commoning different types of property and focuses on 'the suppressed praxis of the commons in its manifold particularities' (Linebaugh 2008: 19). Whether commons are shrinking or growing becomes an empirical question, not something that is derived from the narrative of capital's need and ability to enclose material and immaterial resources. As we grapple with ways of living on this planet, and particularly with ways of managing an open access resource that has been degraded to the extent that all life on the planet is imperilled, this anti-capitalocentric approach of reading the commons for difference perhaps offers a way of expanding the political options that might be open to us to imagine and enact other possible worlds in the here and now.

Constituting an atmospheric commoning-community

Living in Australia it is easy to despair about the state of climate politics. With the election of a conservative government in September 2013, Australia's carbon pricing mechanism (introduced some two years earlier by a Labor government) was not going to last long. When it was repealed in July 2014, the deputy leader of the Greens Party, Adam Bandt, described it as 'the Australian Parliament's asbes-tos moment, our tobacco moment – when we knew what we were doing was harmful, but went ahead and did it anyway' (cited in Baird 2014). A few months earlier at a talk to the Royal Academy, Copenhagen, Bruno Latour had famously referred to the 'Australian strategy of voluntary sleepwalking toward catastrophe' (Latour 2014: 1). When we condense the course of political change to the ups and downs of the election cycle or to the ups and downs of international negotiations on levels of greenhouse gas emissions (GGEs), it is difficult to discern that progress towards constituting an atmospheric commoning-community is being made. By using a different temporality, however, a sense of not only possibility but momentum comes to the fore.

One way of tracking the trajectory of change is to use an intergenerational Commons Yardstick (see Figure 12.1). This simple device helps us to record our relationships to commons over a time period that places our present in the kind of temporal context that climate change requires us to consider (Gibson-Graham *et al.*

A COMMONS YARDSTICK

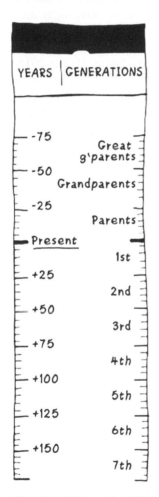

1 GENERATION = 25 YEARS

Figure 12.1 A commons yardstick.

Source: Adapted from Gibson-Graham *et al.* (2013).

2013: 138–9). On it we can record the actions of enclosure or commoning or neglect that have, in previous generations, contributed to current levels of well-being for some and disadvantage for others. And we can look forward, not just to our children's life span but that of seven generations hence, guided by the planning philosophy practised by many Indigenous societies. Taking a generation to be twenty-five years, we can locate commoning (and uncommoning) activities in the past and present, and attempt to forecast 150 years into the future.

This strategy of looking to the past may seem out-of-step with commentators such as Plumwood (2007) who propose that if humans are to survive the Anthropocene we need a different mode of humanity. In an introduction to a special issue on socioecological transformation, the geographer Bruce Braun suggests that 'in the Anthropocene the shape of things to come is increasingly seen to be nonanalogous with what existed in the past' (2015: 239). Braun throws into question the long-standing and strongly-held assumption that 'by understanding the past we might be able to anticipate and shape the future' (Braun 2015: 239). But what if we look to the past with fresh eyes? What if we were to take seriously De Angelis and Harvie's (2013: 292) suggestion that commons may be 'part of a different historical trajectory' than capitalism? This would mean seeing the commons not as subject to the gravitational pull of capitalism nor aligned with a particular form of property. In this section, we use the Commons Yardstick to help identify some of the faltering ways that communities in Australia have, in the past and today, managed to common the atmosphere, working at varying scales and using different strategies. These communities are forging ways of acting as a multi-species planetary community of those who, while they seem to have nothing in common, share the bodily need for clean and coolish air to thrive. The examples we discuss are but one glimpse into practices of commoning that an anti-capitalocentric formulation of the commons brings to visibility. Taken together these examples give insight into a politics appropriate for these climate changing times.

Commoning a city's air

Since being established in 1804, the industrial city of Newcastle, north of Sydney, has relied on large reserves of black coal, initially mined from the immediate city area and more recently from its hinterland. For around 125 years or five generations, dust and smoke-laden air was not just tolerated but embraced as a symbol of the city's fortunes (Bridgman and Cushing 2015: 46). From the late 1930s, however, residents began to express concerns about air pollution and the impacts on their health and the urban environment more generally. People were starting to experience air pollution as something that negatively impacted their lives and well-being. Both white collar and blue collar workers were, in Bruno Latour's terms, 'learning to be affected' by the quality of the air around them (Latour 2004); they were learning that the polluted air was impacting their lungs, their sinuses, their bloodstreams, and they were no longer willing to trade-off their health for economic advancement. At the same time, in 1938, the small local councils of the area were amalgamated into the Council of the City of Greater Newcastle and this meant there was now an authority 'with a much wider geographical brief and sufficient substance to stand up to the large industries' (Bridgman and Cushing 2015: 58). In 1947, in the aftermath of the Second World War, the council established a Smoke Abatement Panel, the first of its kind in Australia (Bridgman and Cushing 2015: 61). The larger council, with its ability to capitalise on advances in technology, installed a network of new dust deposition gauges similar to those that had been used in several British and North American cities. This not only provided a way of quantifying and monitoring what people were experiencing but, because the devices were being used overseas, it enabled comparison with air pollution levels elsewhere. The devices provided 'hard evidence of the extent of the problem' and as a result 'pollution could be less easily ignored' (Bridgman and Cushing 2015: 66). Other developments beyond the city also played a role. Following London's deadly smog of 1952, and the 12,000 deaths that have been associated with the event (Bell and Davis 2001), the New South Wales state government formed a Smoke Abatement Committee in 1955 and introduced the *Clean Air Act* in 1961. These state-level changes combined with the ongoing efforts of the Council and resident groups, supported by local media, meant that by the end of the 1960s, or within one generation of the Council's Smoke Abatement Panel being formed, the average dust fall was reduced from 40.34 tons per square mile per month to 11.47 tons. Newcastle's atmosphere had been commoned and 'the once smoke-obscured cityscape was coming back into focus' (Bridgman and Cushing 2015: 75).

The commoning action involved residents, across class lines, learning to be affected and acknowledging the impact of the embodied experience of living in a highly polluted environment. It also involved shifts in institutional arrangements and the mobilisation of technological advances. It drew on developments locally and those further afield. As a result, a loose community emerged who were able to take responsibility and care for Newcastle's atmosphere, to manage how the atmosphere was being used, and to ensure that there was access to clean air and that this benefit was shared across the city. Air pollution had been a feature of life in Newcastle from the first days of white settlement, and it had been experienced for around five generations before a commoning action was initiated. Once the Second World War had passed, it took only one generation before there was substantial improvement in the air quality. Of course, this is not to say that commoning actions stopped in the 1960s. The struggle to maintain and improve air quality in Newcastle has continued. Most recently it has focused on the health effects of using open coal trains to transport coal from the Hunter Valley to the Port of Newcastle, one of the world's largest coal export ports (to say nothing of the health effects for the planet of burning the coal that is exported from the port).

Commoning the global ozone layer

In addition to this example of a city-level action to common the atmosphere we can also turn to a global-level action. For almost three generations, from the

1930s to the 1970s, chlorofluorocarbons (CFCs) and other ozone-depleting chemicals (ODCs) were manufactured and used without concerns being raised. Then in 1974, in a paper published in *Nature*, two scientists hypothesised that CFCs were migrating into the upper atmosphere and significantly depleting the ozone layer. Public officials in the USA responded and within a few years CFCs were banned in the USA. Once 'the hole' (as it became known) in the ozone layer over Antarctica was detected in 1985 the Montreal Protocol on Substances that Deplete the Ozone Layer was agreed to and a worldwide phase-out of ODCs became a reality. The Montreal Protocol represents, in essence, a national and international community that was constituted to common an unmanaged resource, the ozone layer of the atmosphere. Through the Protocol nations accept responsibility to care for and manage use of ozone so that all species on the planet have access to and benefit from an intact and protected atmosphere (to express the activities of the community in terms of Tables 12.1 and 12.2). The Protocol has been remarkably successful. By 2005 all 191 countries that ratified the Protocol had cut the production and consumption of ODCs by 95 per cent. ODCs have an atmospheric lifetime of 50–100 years so it will still take few generations to repair the damage. Nevertheless the Montreal Protocol has been dubbed not just 'the world's most successful environmental agreement' but also 'the world's most effective climate treaty', given that it has also reduced greenhouse gas emissions 'by the equivalent of approximately 11 gigatons of carbon dioxide a year between 1990 and 2010 … thereby delaying the onset of climate change by up to 12 years' (Grabiel 2007: 20).

Behind this story of successful regulation are multiple stories of the struggle to constitute a community that commons. When we narrate these histories what emerges is a more heterogenous community than that of the nation states and their public officials who were able to formulate an international agreement. Instead we gain insight into the processes of constituting the diverse human and more-than-human community that was needed to common something the scale of the planet's ozone layer. In what follows we can only narrate a few of the factors that played a role in that process, enough, we hope, to provide some understanding of what might be involved in a politics of commoning the atmosphere in the face of the climate crisis.[4]

One factor is that key people were 'learning to be affected'. This is evident from the moment that Professor F. Sherwood Rowland and his postdoctoral fellow Mario J. Molina published their hypothesis in *Nature*. The chemical industry immediately questioned the scientific validity of the research. Du Pont, the world's largest manufacturer of CFCs, even ran a one-page ad in the *New York Times* to make their position public. Rowland and Molina were not dissuaded. Rowland is reported to have said to his wife: 'The work is going well. But it looks like the end of the world' (cited in Roan 2012). This concern drove him (and Molina) to continue with their research and to speak-out publically to raise the alarm. We could say that Rowland and Molina were learning to be affected by the chemical reaction that occurs when CFCs reach the stratosphere and break down to release chlorine atoms. Politicians and policy-makers in the

USA were also learning to be affected and they responded to the call by taking early action well before the science of ozone depletion was settled. From 1977 these public officials tried to persuade other nations to take action to ban CFCs with little success until a breakthrough in 1985.

This breakthrough was the result of a second factor that played a role in constituting a commoning-community – the contribution of technological advances. It took two research teams using both long-standing and new technologies to make the breakthrough. Scientists from the British Antarctic Survey had detected ozone depletion in Antarctica using ground-based instruments such as the Dobson Ozone Spectrophotometer (a device developed in the mid-1920s). They reported their findings in *Nature* in May 1985. At the same time, NASA scientists had been tracking ozone in the atmosphere using satellite-based sensors and they were able to verify the extent of ozone depletion, and produce the first visual images of the so-called hole in the ozone layer. These images were presented in August 1985 at a conference in Prague by NASA scientist Dr Pawan Bhartia who recounts, 'All hell broke loose, particularly in the media. People were scared and thought this could be a real disaster that could kill us, give us cancer' (Hansen 2012). One result was that nations reacted quickly with the Montreal Protocol being agreed to within two years, in September 1987 (and coming into force from January 1989).

Australia's geographic proximity to the ozone hole was a third factor that helped to constitute a commoning-community. Its geographic proximity to the ozone hole means that it is particularly vulnerable. Ozone depleted air travels from Antarctica to Southern Chile and Southern Argentina, New Zealand and Australia. The main human effect is that by letting in more ultraviolet B (UV-B) radiation there is an increase in skin cancers and cataracts, a particular concern in Australia which has the world's highest rate of skin cancer.[5] Since the early 1980s public health authorities had been raising awareness of the importance of sun protection. When the images of the hole in the ozone layer were presented in 1985 the Australian public were primed to react. Thanks to the sun protection campaigns they were already learning to be affected by UV radiation. A hole in the ozone layer that would increase their vulnerability was unacceptable and they urged their government to act (Andersen and Sarma 2002: 208). This pressure contributed to the Australian government playing an influential role in formulating the terms of the Montreal Protocol. In particular Australia 'often served as a bridge between the concerns of developed and developing countries' (Andersen and Tope 2002: 255), and this helped with what are generally acknowledged as two of the most important features of the Protocol: the different deadlines for phase-out for developed and developing countries, and funding support to help developing countries meet their compliance targets.[6]

The concerns expressed by the Australian public as they were learning to be affected also helped to create a context in which others were prepared to take action. For example, the Australian Plumbers and Gasfitters Employees Union, prompted by public concern, refused to install new firefighting systems based on halon (one of the ODCs) when alternatives were available. They also agreed to maintain existing halon-based systems only under certain conditions (Andersen and Tope 2002). The debate that ensued resulted in the fire protection industry in Australia, in 1989, abandoning its strong opposition to controls on halon, and internationally had the effect of prompting the industry to be pro-active in developing alternatives. This type of action helped to pressure the chemical industry (including Du Pont) to devise non-ozone destroying substitutes.

What we can see happening is that a number of factors came together to give impetus to the formation of a community. There is no doubt that scientific and technological advances, especially in satellite-based remote sensing and computer firepower, helped to identify the problem. That there was 'a hole' in the ozone that could be captured in a media-grabbing image and phrase meant that the problem could be easily and powerfully communicated. Indeed, we could even say that it was fortuitous that the chemical reaction occurs seasonally and variably over Antarctica; each Southern Hemisphere spring there is intense media reporting about the size of the hole in the ozone. Is it getting bigger? Is it getting smaller? This cyclical event makes headlines. A more steady and uniform thinning would perhaps be less newsworthy. Alongside these factors, people were learning to be affected by the chemical reaction in various ways and this helped contribute to a successful international institutional agreement. As a result the community that formed to common the ozone layer through the Montreal Protocol included more than nation states and their public officials. It included scientists, unionists, multinational firms, and media reporters, and it included more-than-human elements such as chemical compounds and reactions, satellites and scientific instruments, and media images and maps – to name just some of the elements that entangled as part of the community that commons.

There seems to be a huge gap between the slow progress of negotiations under the United Nations Framework Convention on Climate Change and the achievements of the Montreal Protocol. Might thinking about the process of constituting the heterogeneous community that commons help us find grounds for hope?

Creating a solar commons

Despite all that is known about greenhouse gas emissions and global warming, Australia continues to voraciously export and burn black and brown coal which, by the geological luck of the draw, it has in copious quantities. During the twentieth century, national development was built on the provision of cheap coal-fired electricity delivered to business and residential consumers by state government owned monopolies (Gibson 2001). But as we move into the twenty-first century there are indications that change is afoot and that a nascent community that commons is starting to form with the energy of the sun at its centre.

Along with coal, Australia is also blessed with abundant sunlight. Australian researchers have been at the forefront of developing solar technologies for three to four generations, starting with solar water heaters in 1950s and photovoltaics in the 1970s. Research teams such as those in the School of Photovoltaic and

Renewable Energy Engineering at the University of New South Wales have been responsible for breakthroughs in photovoltaic technology and for training several generations of solar innovators who are now based in Australia and overseas.[7] From the 2000s, or within the current generation, there has been a massive uptake of these solar technologies at the household level, including in low-income households (Green Energy Trading 2014). As at August 2015, there were over 1.4 million small-scale solar panel systems installed on Australian rooftops and a further 935,000 solar hot water systems (Australian Government Clean Energy Regulator 2015).[8] The Australian Energy Market Operator (2011: 9–2) has been reporting that the rapid uptake of solar at the household level is leading to an unprecedented reduction in electricity demand. Their prediction is that with consumers increasingly keen to minimise their environmental footprint the current electricity generation industry is being confronted not with incremental change but with 'an abrupt "step change"' (2011: 9–1). This grass-roots movement at the household level is reshaping the energy sector, and making it not inconceivable that the coal-fired power stations and the poles and wires of the electricity grid will be obsolete in the not too distant future.

The uptake of household solar has been driven by a number of motivations and incentives. As the Australian Energy Market Operator has identified, people are keen to minimise their environmental footprint. We could say that this is a result of people learning to be affected by the impacts of climate change; in Australia this means being affected by the embodied experience of hotter and drier summers, and more extreme climate events such as cyclones (see also Cameron 2011). Such experiences are prompting citizens to consider how their everyday practices come to bear on these changes and how, through the adoption of technologies such as solar power, they can play a role in slowing the future that is pressing in. In previous research we also found that frustration with the lack of concerted government action on climate change (such as Australia's refusal to ratify the Kyoto Protocol until the election of a Labor Federal Government in 2007) provided impetus for citizens to take action in various ways including, for example, by participating in a community-run scheme for the bulk purchase and installation of household solar panels via a cooperative solar business (Cameron and Hicks 2014). Such motivations have been helpfully realised by financial incentives such as the federal government's cash rebate of up to $8,000 for the installation of solar panels (through the Solar Homes and Communities Programme, 2000–09) and up to $1,600 for the installation of solar water heaters (through the Energy Efficient Homes Package, 2009–10).[9] State-based feed-in tariffs offer income generating opportunities for households producing excess power from their solar panels. There are also financial innovations, including solar leasing and power purchase agreements, that are experimenting with financing small-scale and distributed systems rather than the centralised arrangements of old.

The movement Solar Citizens is one public face of the rapidly growing number of solar households. With a membership of over 70,000, Solar Citizens aims to be a voice for 'solar owners and supporters' (see www.solarcitizens.org.

au/about_us). Its current 'Stand up for solar' campaign includes lobbying for 50 per cent of Australia's energy to come from renewable sources by 2030; a solar ombudsman; and a national programme to help low-income households and renters to install solar power. But behind this public face lies the million plus households who are consumer-producers or 'prosumers' of solar power and solar hot water, and who are increasingly seen as a political force to be reckoned with. 'The solar vote' has become a feature in all recent elections and by-elections: there are interactive maps by electorate showing the number and proportion of homes with solar energy; lobby groups produce 'solar scorecards' for voters with information on the solar policies of the major political parties; and there is analysis of how the solar vote is likely to affect outcomes. A new political constituency of solar prosumers and solar supporters is emerging.

A community that is commoning the energy of the sun is being created. In turn this community is contributing to a larger community that is commoning the atmosphere by finding ways of reducing and managing GGEs. In terms of the Ways of Commoning diagram (Table 12.2), this community is not so much commoning an unmanaged resource as creating and commoning a resource, solar power, out of something whose potential has not been realised, the sun. This community is helping to share and widen access to solar power, and through the care of initiatives such as Solar Citizens and their campaign for a solar ombudsman to ensure that the 'industry' operates in a well-managed and responsible way so that the benefits of solar power can be realized. Solar Citizens is one human face of this emerging community, but it is a heterogeneous community that includes the existing and emerging technologies of solar capture and storage, the roofs of houses, the household smart meters that help monitor and manage energy use; new financing arrangements; and a market that has been constituted partly by the intervention of government subsidies and feed-in tariffs. This combination of factors has had cascading effects helping create an opening for solar power and driving further technical innovation, reducing the price, and increasing the efficiency of solar technologies, making the technology more widespread – and the current coal-based system of electricity generation more obsolete.[10] This emerging solar commoning-community is but one example of an energy-based commons. In Australia there is also a new commoning-community emerging in the area of community-owned wind farms (Cameron and Hicks 2014).[11] When we look overseas there is likewise a proliferation of energy-commoning-communities. It may be that these initiatives, as much as international agreements, are key to action on climate change.

The practice of commoning as a postcapitalist politics

The current and pressing need for rapid social and ecological transformation calls for unprecedented action at all scales. In this chapter we have proposed that commoning might become the focus of a politics for our times. We have shown how, over almost four generations, different commoning-communities have formed to care for the atmosphere in different ways. The question is, can we accelerate action to cool our warming atmosphere in just one generation? And what role could we as social scientists play in 'transformations for sustainability'?[12]

We have argued that the current debate about commons and commoning is still largely beholden to a capitalocentric discourse and that this framing stands in the way of a more concerted application of transformative commoning strategies. In the capitalocentric framing, the politics that emerges remains linked to an abstract and vague multitude – the coming community of commoners. When this framing is linked to the understanding of commons as a form of property then politics focuses on struggles against enclosure and privatisation. This is not to say that these struggles are not important; however, our concern is with a mode of politics that can respond now to calls for a different mode of humanity. In this new mode, humans might take their place as only one in a multi-species community of life on this planet, abandoning illusions of mastery to become 'team players' with non-human earth others. Might the anti-capitalocentric understanding of commoning that we have presented in this chapter be the basis for this type of politics? This would be a transformative politics not located in an elusive future, nor in a combative anti-privatisation struggle focused narrowly on property rights; instead it would be a postcapitalist and posthuman politics located in the shared present – in the becoming of a commoning-community in the here and now.

To help build this politics we have taken seriously the suggestion that the commons might be part of a different historical trajectory. We have identified moments when commoning has been successful and unpicked some of the constituents of those commoning-communities that have effected change on city, national, and global scales. Our re-narrativisation has expanded our understanding of the community that commons and produced a sense of the diverse temporalities of transformation.

What we have learnt is that the commoning-community is more-than-human. The agent of change, the commoner, is no longer (and perhaps never was) a person or a category such as the working class but an assemblage.[13] Certainly these assemblages include humans, but they also include non-humans; they may include class but also non-class alignments; they may include social movements and grass-roots organisations but also governments, institutions, and firms; they may include non-market mechanisms but also markets; they may include animate beings who have nothing in common except breathing and living, but also inanimate entities that share an existence on this planet. As social scientists we have a role to play in helping to identify the human and more-than-human actants of the commoning-community. This may involve working with technologists, scientists, biota, and so on to enrol members of the commoning-community. Our work is to help forge connections between things (as we've done in the three examples in this chapter).

We have also learnt that the commoning-community is not always recognisable as a community, even to itself. The community that commons is perhaps more easy to detect in retrospect. Thus, in a climate changing world, where the window for meaningful action is rapidly diminishing, as social scientists our work is to seek out those nascent connections and associations that will help construct emerging commoning-communities. This is not a matter of making more visible something that pre-exists discourse, but of proposing and performing commoning-communities as a means of strengthening embryonic communities and engendering more. We know, for example, that there are a multitude of energy-based commoning-communities already operating across the globe. Might these endeavours be further mobilised by making the commoning assemblage they are part of more apparent and encouraging this decentralised punctiform place-based politics to mesh as a global network? The work of this assemblage is already being done – it is creating energy resilience, and markets are already communicating via pricing signals that a new mode of care for the air has arrived. What is lacking is a name for this politics; perhaps 'commoning' is one that could enrol a wide spectrum of support.

This form of language politics could be one contribution to helping sustain the emerging solar commons. Here it is important to recognise that this language politics also involves renaming or reframing. For example, financial capital is involved in solar commoning through the development of financial instruments that enable small-scale and decentralised power generation; productive capital is involved through the development of the technology. The temptation is to read this as simply the expansion of capital-centred activities and the colonisation of new areas of social and economic life. However, we might also reframe these activities by saying that capitalist activities are being drawn into, perhaps even co-opted, as part of the postcapitalist politics of commoning. In other words, the postcapitalist project is not necessarily anti-capitalist. In the same way that James Ferguson argues that neoliberal ideas and techniques do not have to result in neoliberal outcomes and that any governmental technique is accompanied by 'radical political indeterminacy' (2015: 31), so too capitalist economic activities do not have to result in capitalism as we know it as the outcome (see also Gibson-Graham 1996, 2006). Capitalist economic activities are accompanied by 'radical economic indeterminacy'. Furthermore, as part of a commoning assemblage, financial and productive capital also play a role in helping to sustain the emerging solar commons for as the financial instruments and technologies become embedded and integrated into the solar commons they help to make the assemblage more durable.

Finally, we have learnt that commoning, while it might be ever present, marches to an irregular beat. Commoning is a messy and fragmented process in which transformation takes place with different rhythms over a long timeframe. Some things seems to happen very quickly such as the rapid take up of solar energy in Australia in the last few years or the two years that it took to get the Montreal Protocol signed. But if we focus only on these moments we miss the work that can go on for generations to help create the conditions for what seems like rapid transformation. This is not to say that change happens in a linear or predictable way. The process of change emerges out of any number of things coming together and entangling to create the conditions for shifts to occur. Only with an inter-generational perspective can we begin to see the multiple temporalities at work and by which change takes place. Our work is to interrogate and listen differently so we might see how to accelerate the pace of change while also being attentive to ways of working with different temporalities, including those of the more-than-human world. At present we are confronted with a ticking climate time bomb; we need to become acquainted with the ways that society can change suddenly without wishing upon ourselves the equivalent of more 'readjustment events' such as the Lisbon earthquake of 1755 (Connolly 2013), the Sumatra-Andaman earthquake of 2004 (Clark 2011), or the Christchurch earthquake of 2011 to prompt radical transformation. Our efforts in this chapter are one step towards strengthening our ability to see ourselves as part of a commoning-community assemblage that acts, and that must keep acting to care for, take responsibility for, and ultimately share the benefits of life with all earth others.

Notes

1 We would like to thank Ash Amin and Philip Howell for organising the Shrinking Commons Symposium at Cambridge University in 2014 and this subsequent volume. We acknowledge support from the Julie Graham Community Economies Research Fund to attend the Community Economies Writing Retreat in Bolsena, Italy in August 2015 during which Katherine Gibson's Symposium presentation was significantly expanded into the current chapter. We also acknowledge the direct and indirect contribution of the knowledge commons, that is, the Community Economies Collective.

2 As we have discussed elsewhere, the effect of this widespread mode of theorising denies forms of economic difference any independent identity, effectivity, or dynamism (e.g. Gibson-Graham 1996, 2006).

3 We also acknowledge that Linebaugh identifies that commoning can be co-opted by, in his words, 'capitalists and the World Bank' (2008: 279).

4 Examples of more comprehensive accounts of the process of formulating the Montreal Protocol include Andersen and Sarma (2002) and Grundmann (2001).

5 Other effects include diminished yields from agricultural crops and loss of phytoplankton with implications for the health of entire marine ecosystems.

6 For example, halons were phased out in developed countries by 1993 and by 2010 in developing countries; CFCs were phased out by 1995 in developed countries and by 2010 in developing countries.

7 The alumni include Andrew Birch, co-founder and CEO of Sungevity Inc. based in the USA, and Shi Zhenrong, founder of Suntech Power, based in China.

8 This is in a country with approximately 9.1 million dwellings (Australian Bureau of Statistics 2013). It is also worth noting that just ten years ago in 2005 there were only around 3,500 small-scale solar panel systems and 125,000 solar hot water systems (Clean Energy Regulator 2015).

9 The Energy Efficient Homes Package was part of the Nation Building Economic Stimulus Plan, the federal government's initiative to address the global financial crises.

10 The cost of the average installed solar PV system in 2012 was around one quarter of the cost in 2002. In that ten-year period the cost dropped from around $13 per watt to $3 per watt (Flannery and Sahajwalla 2013: 23).

11 Anecdotally, there is evidence that comments in early 2015 by the then Australian Prime Minister and Treasurer about the ugliness of wind turbines and the launch of a parliamentary inquiry into the health effects of wind farms (while at the same time the conservative federal government supported and defended coal mining) had the effect of increasing public support for wind energy and invigorating people's interest in the potential of community-owned wind farms.

12 Our use of this phrase deliberately echoes the name of the International Council of Social Sciences programme which has social scientists as the lead scientists in cross-disciplinary teams that are researching ways of accelerating social change to address problems of global change and sustainability.

13 This is an area of key thinking in the Community Economies Collective. For example, see Gibson-Graham and Miller (2015), Hill (2015), Roelvink (2013), and Roelvink *et al.* (2015).

References

Andersen, Stephen O. and K. Madhava Sarma. *Protecting the Ozone Layer: The United Nations History*. London: Earthscan, 2002.

Andersen, Stephen O. and Helen K. Tope. 'The Importance of Australian Leadership in the Montreal Protocol', in Stephen O. Andersen and K. Madhava Sarma (eds), *Protecting the Ozone Layer: The United Nations History*. London: Earthscan, 2002, 255–6.

Australian Bureau of Statistics. *Perspectives on Regional Australia: Housing Arrangements – Home Ownership in Local Government Areas, 2011*. 2013, at: www.abs.gov.au/ausstats/abs@.nsf/mf/1380.0.55.010?OpenDocument (accessed 18 January 2016).

Australian Energy Market Operator. 'National Transmission Network Development Plan for the National Energy Market, AEMO'. 2011, at: www.aemo.com.au/Electricity/Planning/Archive-of-previous-Planning-reports/2011-National-Transmission-Network-Development-Plan/~/media/Files/Other/planning/ntndp/NTNDP2011_CD/documents/NTNDP_2011%20pdf.ashx (accessed 18 January 2016).

Australian Government Clean Energy Regulator. *Postcode Data for Small-scale Installations*. 8 September 2015, at: www.cleanenergyregulator.gov.au/RET/Forms-and-resources/Postcode-data-for-small-scale-installations#Smallscale-installations-by-installation-year (accessed 18 January 2016).

Baird, Julia. 'A Carbon Tax's Ignoble End: Why Tony Abbott Axed Australia's Carbon Tax'. *New York Times* 24 July 2014, at: www.nytimes.com/2014/07/25/opinion/julia-baird-why-tony-abbott-axed-australias-carbon-tax.html (accessed 18 January 2016).

Bell, Michelle L. and Devra Lee Davis. 'Reassessment of the Lethal London Fog of 1952: Novel Indicators of Acute and Chronic Consequences of Acute Exposure to Air Pollution'. *Environmental Health Perspectives* 109, Supplement 3 (2001): 389–94.

Braun, Bruce. 'Futures: Imagining Socioecological Transformation – An Introduction'. *Annals of the Association of American Geographers* 105, no. 2 (2015): 39–43.

Bresnihan Patrick and Michael Byrne. 'Escape into the City: Everyday Practices of Commoning and the Production of Urban Space in Dublin'. *Antipode* 47, no. 1 (2015): 36–54.

Bridgman, Howard and Nancy Cushing. *Smoky City: A History of Air Pollution in Newcastle, NSW*. Hamilton, NSW: Hunter Press, 2015.

Cameron, Jenny and Jarra Hicks. 'Performative Research for a Climate Politics of Hope: Rethinking Geographic Scale, "Impact" Scale and Markets'. *Antipode* 46, no. 1 (2014): 53–71.

Cameron, Jenny, with Craig Manhood and Jamie Pomfrett. 'Bodily Learning for a (Climate) Changing World: Registering Differences through Performative and Collective Research'. *Local Environment* 16, no. 6 (2011): 493–508.

Chakrabarty, Dipesh. 'The Climate of History: Four Theses'. *Critical Inquiry* 35, Winter (2009): 197–222.

Clark, Nigel. *Inhuman Nature: Sociable Life on a Dynamic Planet*. London: Sage, 2011.

Clean Energy Regulator (Australian Government). 'Postcode Data for Small-scale installations', 2015, at: www.cleanenergyregulator.gov.au/RET/Forms-and-resources/Postcode-data-for-small-scale-installations#Smallscale-installations-by-installation-year (accessed 18 January 2016).

Climate Commission. 'The Critical Decade: Australia's Future – Solar Energy'. Climate Commission Secretariat, Department of Industry, Innovation, Climate Change, Science, Research and Tertiary Education, 2013, at: www.climatecouncil.org.au/uploads/497bc d1f058be45028e3df9d020ed561.pdf (accessed 18 January 2016).

Connolly, William E. *The Fragility of Things: Self-Organizing Processes, Neoliberal Fantasies, and Democratic Activism*. Durham, NC: Duke University Press, 2013.

De Angelis, Massimo. 'On the Commons: A Public Interview with Massimo De Angelis and Stavros Stavrides'. *e-flux* 17, June–August (2010): 1–17.

De Angelis, Massimo and David Harvie. 'The Commons', in Martin Parker, George Cheney, Valérie Fournier and Chris Land (eds), *The Routledge Companion to Alternative Organization*. London: Routledge, 2013, 280–94.

Ferguson, James. *Give a Man a Fish: Reflections on the New Politics of Distribution*. Durham, NC: Duke University Press, 2015.

Flannery, Tim and Veena Sahajwalla. *The Critical Decade: Australia's Future – Solar Energy*. (Australian Government) Climate Commission Secretariat, Department of Industry, Innovation, Climate Change, Science, Research and Tertiary Education, 2013, at: www.climatecouncil.org.au/uploads/497bcd1f058be45028e3df9d020ed561.pdf (accessed 18 January 2016).

Gibson, Katherine. 'Regional Subjection and Becoming'. *Environment and Planning D: Society and Space* 19, no. 6 (2001): 639–67.

Gibson-Graham, J.K. *The End of Capitalism (As We Knew It): A Feminist Critique of Political Economy*. Oxford: Blackwell, 1996.

Gibson-Graham, J.K. *A Postcapitalist Politics*. Minneapolis, MN: University of Minnesota Press, 2006.

Gibson-Graham, J.K. and Ethan Miller. 'Economy as Ecological Livelihood', in Katherine Gibson, Deborah Bird Rose, and Ruth Fincher (eds), *Manifesto for Living in the Anthropocene*. New York: Punctum Books, 2015, 7–16.

Gibson-Graham, J.K., Jenny Cameron, and Stephen Healy. *Take Back the Economy: An Ethical Guide for Transforming our Communities*. Minneapolis, MN: University of Minnesota Press, 2013.

Grabiel, Danielle Fest. 'Crucial Crossroads'. *Our Planet*, Special Issue, *Celebrating Twenty Years of the Montreal Protocol*, September (2007): 20–1.

Green Energy Trading. *Postcode and Income Distribution of Solar*. Report for the REC (Renewable Energy Certificates) Agents Association, April 2014, at: www.recagents. asn.au/wp-content/uploads/2014/04/GET-Postcode-report-for-RAA-April-2014.pdf (accessed 18 January 2016).

Grundmann, Reiner. *Transnational Environmental Policy: Reconstructing Ozone*. London: Routledge, 2001.

Gudeman, Stephen. *The Anthropology of Economy: Commodity, Market and Culture*. Oxford: Blackwell, 2001.

Hansen, Kathryn. 'Discovering the Ozone Hole: Q&A with Pawan Bhartia'. NASA Feature, 17 September 2012, at: www.nasa.gov/topics/earth/features/bhartia-qa.html (accessed 18 January 2016).

Hardin, Garrett. 'The Tragedy of the Commons'. *Science* 162, no. 3859 (1968): 1243–8.

Hardin, Garrett. 'Extensions of "The Tragedy of the Commons"'. *Science* 280, no. 5364 (1998): 682–3.

Hardt, Michael. 'The Common in Communism'. *Rethinking Marxism* 22, no. 3 (2010): 346–56.

Hardt, Michael and Antonio Negri. *Commonwealth.* Cambridge, MA: Harvard University Press, 2009.

Harvey, David. *The New Imperialism.* Oxford: Oxford University Press, 2003.

Hill, Ann. 'How Commoning Happens: Learning from the Philippines'. Unpublished, available from the author, 2015.

Huron, Amanda. 'Working with Strangers in Saturated Space: Reclaiming and Maintaining the Urban Commons'. *Antipode* 47, no. 4 (2015): 963–79.

Latour, Bruno. 'How to Talk about the Body? The Normative Dimension of Science Studies'. *Body and Society* 10, no. 2/3 (2004): 205–29.

Latour, Bruno. 'On Some of the Affects of Capitalism'. Lecture Given at the Royal Academy, Copenhagen, 26 February 2014.

Linebaugh, Peter. *The Magna Carta Manifesto: Liberties and Commons for All.* Berkeley, CA: University of California Press, 2008.

Nancy, Jean-Luc. 'Of Being-in-Common'. Translated by James Creech, in Miami Theory Collective (eds), *Community at Loose Ends.* Minneapolis, MN: University of Minnesota Press, 1991, 1–12.

Ostrom, Elinor. *Governing the Commons: The Evolution of Institutions for Collective Action.* Cambridge: Cambridge University Press, 1990.

Plumwood Val. 'A Review of Deborah Bird Rose's "Reports from a Wild Country: Ethics of Decolonisation"'. *Australian Humanities Review* 42, August (2007): 1–4.

Roan, Shari. 'F. Sherwood Rowland Dies at 84; UC Irvine Professor Won Nobel Prize'. *Los Angeles Times* 12 March 2012, at: http://articles.latimes.com/2012/mar/12/local/la-me-sherwood-rowland-20120312 (accessed 18 January 2016).

Roelvink, Gerda. 'Rethinking Species-Being in the Anthropocene'. *Rethinking Marxism* 25, no. 1 (2013): 52–69.

Roelvink, Gerda, Kevin St Martin, and J.K. Gibson-Graham (eds). *Making Other Worlds Possible: Performing Diverse Economies.* Minneapolis, MN: University of Minnesota Press, 2015.

St Martin, Kevin. 'Disrupting Enclosure in New England Fisheries'. *Capitalism Nature Socialism* 16, no. 1 (2005): 63–80.

Index

Page numbers in *italics* denote tables, those in **bold** denote figures.

Taylor & Francis eBooks

Helping you to choose the right eBooks for your Library

Add Routledge titles to your library's digital collection today. Taylor and Francis ebooks contains over 50,000 titles in the Humanities, Social Sciences, Behavioural Sciences, Built Environment and Law.

Choose from a range of subject packages or create your own!

Benefits for you

» Free MARC records
» COUNTER-compliant usage statistics
» Flexible purchase and pricing options
» All titles DRM-free.

Benefits for your user

» Off-site, anytime access via Athens or referring URL
» Print or copy pages or chapters
» Full content search
» Bookmark, highlight and annotate text
» Access to thousands of pages of quality research at the click of a button.

Free Trials Available
We offer free trials to qualifying academic, corporate and government customers.

eCollections – Choose from over 30 subject eCollections, including:

Archaeology	Language Learning
Architecture	Law
Asian Studies	Literature
Business & Management	Media & Communication
Classical Studies	Middle East Studies
Construction	Music
Creative & Media Arts	Philosophy
Criminology & Criminal Justice	Planning
Economics	Politics
Education	Psychology & Mental Health
Energy	Religion
Engineering	Security
English Language & Linguistics	Social Work
Environment & Sustainability	Sociology
Geography	Sport
Health Studies	Theatre & Performance
History	Tourism, Hospitality & Events

For more information, pricing enquiries or to order a free trial, please contact your local sales team:
www.tandfebooks.com/page/sales

 Routledge Taylor & Francis Group | The home of Routledge books | **www.tandfebooks.com**

Milton Keynes UK
Ingram Content Group UK Ltd.
UKHW031148141024
449569UK00024B/976